2017 EDITION

The

idealw

# FIELD GUIDE

## to Software
## for Nonprofits

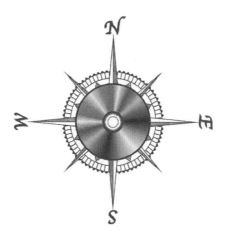

A Quick Guide to Essential
Software for your Organization

THE FIELD GUIDE TO SOFTWARE

Welcome,

**We created The Field Guide to Software** to help pinpoint the types of software that might be useful for the needs of nonprofits. It's been a few years since our last update, but this new edition follows the same basic principles. Through high-level overviews and user-friendly summaries, it can demystify all the possible options available to you and get you started on finding the right tools to help you better meet your missions.

What does it cover? Tried-and-true and emerging tools and technologies, best practices, and specific aspects of nonprofit software. There's also a section to guide you through the sometimes daunting process of choosing and implementing software.

We know you have your hands full and don't always have time to keep up with the latest information about technology for your nonprofit. Let us help. Start with this guide, and then explore the wide library of free articles and reports and our training archive at www.idealware.org.

Thank you for all you do to make the world a better place. We hope this Field Guide will help you do it all just a little more easily.

*Karen Graham*

**Karen Graham**
Executive Director, Idealware

THE FIELD GUIDE TO SOFTWARE

TABLE OF CONTENTS

# 5. Collaboration

# 6. Constituent Management

# 7. Fundraising and Events

*What types of software would be helpful for your nonprofit in the areas of back office and productivity, analytics, collaboration, constituent management, fundraising and events, and communications? The Field Guide is designed to answer exactly that question. This Introduction helps you understand how to use this guide, walks you through the software we believe every nonprofit organization should have, and provides an overview of a few key terms.*

INTRODUCTION

# Understanding What You Need

## Ready to think through the types of software your nonprofit is likely to find helpful? Dive right in...

This Introduction walks you through the software we believe every nonprofit organization should have, regardless of size or mission, and provides an overview of a few key terms that apply to nonprofit software.

The Case Study section provides an overview of the different types of software used by a set of fictional—but plausibly realistic—nonprofits. If you're unsure what you might need, this is a good place to start. These examples can help you pinpoint the specific types of software to investigate further in the next section.

The remainder of the guide is organized to help you find the types of software that might be useful based on the goals you're trying to achieve. We'll walk through the software designed for six different areas: Back Office and Productivity, Analytics, Collaboration, Constituent Management, Fundraising and Events, and Communications. In each section, we'll first take an overall look at the types of software that might be useful based on your own situation and your organization's level of technical sophistication. Each section then contains an introduction to the software types that might be useful in that area.

Note that some systems overlap different areas. For example, Page Layout software can be considered part of both the Back Office and Productivity area and Communications area because of the different ways in which an organization might use it. In those cases, we included the full description of the systems in the area in which we think they fit best for their primary use, but also refer to them in other sections.

The eighth section, Choosing and Implementing Software, walks you through the process of selecting the best tool for your budget and needs and putting it to use at your organization.

We recommend that you begin with the case studies or the walk-throughs for each category to identify the types of software you're likely to find useful. Every time you see a software type **highlighted like this**, it means we covered it elsewhere in this guide, so you can refer to the appropriate section for more information.

Where did all this information come from? Our mission is to provide information to help nonprofits make smart software decisions. Over the years we've done impartial research and reviews of many different types of software. This Field Guide is the synthesis of all that research—it boils thousands of pages of reports and articles down to a handy, concise guide.

# Every Organization Needs

## Systems that help with common tasks—such as managing constituents and maintaining websites—can benefit nearly every organization, regardless of mission or budget.

### Back Office and Productivity

Whatever your mission, there are certain tools you'll need to manage your organization and staff day in and day out. If you have more than one staff member, an Excel spreadsheet will not be sufficient for accounting purposes—you'll need a dedicated *Accounting System* to track finances, expenditures, and payroll.

*Office Software* is important to help you create and edit documents, spreadsheets, presentations, and all the other materials on which organizations run. *Email and Calendar* software lets you and your staff send and receive critical emails and share your schedules. You'll need an internet connection to support email, obviously. And if you have an internet connection, you need *Virus Protection* software to keep malicious computer viruses and spyware from compromising your data or your productivity. *Firewalls* are also important to prevent hackers and others from gaining unauthorized access to your data and computers.

Finally, you'll need a *File Backup and Recovery* solution to protect your organization's data and save you the time, cost, and effort of recovering from a data loss.

## Analytics

You can learn a lot about the impact of your programs and what people think of your work by collecting and analyzing data. To help with **Program Evaluation**, many organizations create **Dashboards** that aggregate constituent data, **Survey Data**, **Website Analytics**, and more. They also use tools for **Measuring Social Media** and **Online Listening**. Together, all of these tools can help you get a clear picture of your organization's effectiveness.

## Collaboration

Whether you have two staff members or two thousand, you need some way to share and manage files with each other and with others outside your organization. A good **File Sharing** system won't break the bank, but will make your life noticeably easier. **Online Chat** and **Online Conferencing** can also help break down barriers.

## Constituent Management

Most fundamentally, you'll need some kind of software to track donors, event attendees, volunteers, and other constituents. Excel spreadsheets are great for maintaining a family task list, but quickly show their weakness when you begin tracking even just a few complex interactions, such as donations. A good database should let you store all the information you'll need about all your supporters.

What type of system will best help you will depend on your specific needs and budget. There are a number of basic types, each ranging widely in cost depending on the features you want. Every organization needs some type of system—**Case Management Software**, **Donor Management System**, **Constituent Relationship Management**, **Volunteer Management Software**, or even a **Specialized Constituent Management System**—but which one makes the most sense for you depends on your specific needs.

## Fundraising and Events

Of course, it's important for every organization to raise funds. To do that well, you'll need some kind of system to track constituents, and effective ways to reach out and engage people, but you might also need specialized tools for *Event Registration*, *Crowdfunding and Peer-to-Peer Fundraising*, or *Online Donations*.

## Communications

In addition to tracking constituents, it's important to have **Website Content** that clearly communicates who you are, what you do, and what value you provide to your community. In order to easily update your website yourself with new information or events as they happen, regardless of how technical you are, you'll need a web **Content Management System** and should make sure that it enables **Mobile-Friendly Websites**.

You should also have a **Broadcast Email** package designed to send out emails to hundreds (or hundreds of thousands) of people. Email is a quick and cost-effective supplement to direct mail or face-to-face communication, and a great way to reach out to or engage constituents or to fundraise, but it's important to have a different tool for your broadcast email than you use for your individual emails.

You also need to create opportunities for two-way conversations. **Social Media** platforms are good places to not only share information, but hear a variety of perspectives and participate in conversations that go far beyond your organization's walls.

To create content that will be interesting and engaging, you'll likely need to think about **Graphics and Multimedia**. **Photo Editing** and **Page Layout** tools can help you present vivid content. There are also many free and low-cost **Multimedia Editing** tools and platforms specifically designed for **Video Sharing and Streaming**.

*How can different types of nonprofits most effectively use software? This section provides examples of how fictional but realistic organizations use software to meet their needs. It starts with smaller organizations that aren't yet ready to make a big investment in software, and then moves to larger organizations with more complex needs.*

*All of the software names highlighted within the text are covered in more detail in this guide—look them up to learn more.*

# Using Technology to Rally Support for Waterways

## The River and Lake Council*

*34 staff/$5.1 million annual budget*

For half a century, The River and Lake Council has worked to protect the waterways of Wisconsin. The organization uses its money people power for habitat restoration, public education programs, and legislative advocacy. "Water is everything to us in Wisconsin," said Steve Drury, Executive Director. "We fish it. We drink it. Our children play in it."

A six-person fundraising team works to raise money through a wide assortment of campaigns, special events, and grant support. The headline event is River Days, a two-week fundraising push held each May that includes a publicity campaign, a "biggest catch" competition, and walk-a-thon, culminating in a gala dinner and auction. The logistics are daunting, but sophisticated software makes managing them possible. "It's a lot to tackle, but fortunately we have good technology," Drury said.

The team uses **Event and Auction Management** software to manage the complicated gala arrangements, including the seating plan and the on-site live auction. **Credit Card Processing** systems let them accept credit cards for auction purchases, saving staff the hassle and extra paperwork of the carbon copy imprinter they used to use. The team also uses an **Event Registration** tool to help sell tickets online. There's an online component, as well, which the team runs using **Online Auction** software.

Every year the River and Lake Council screens a new "Water Workers" video at the gala to showcase the organization's work and to honor its many volunteers. "Everyone works so hard. It's fun to show them off a little," Drury said. Team members used to outsource the video production to professionals. However, in recent years they've begun producing the video in-house using **Multimedia Editing** tools.

A major focus of River Days is the Saturday walk-a-thon along the Mississippi River. Many supporters have participated for years, raising money from their friends and family. For the first time last year, the organization also tried using an online *Peer-to-Peer Fundraising* component for the walk-a-thon, which worked well. Some supporters were more comfortable creating their own online fundraising pages to reach out to people they know, rather than asking them in person. Others enjoyed coming up with their own fundraising strategies for the walk, which they shared in the organization's custom online community. The online community is also where the council's volunteer coordinator posts activities and the communications director shares successes.

Despite the success and popularity of River Days, fundraising and outreach isn't just a two-week affair. Recognizing this, the team mounts various campaigns and appeals throughout the year. A robust *Donor Management* system lets them manage a series of direct mail appeals and a substantial foundation-grant outreach effort. The Council has integrated other individual tools with its *Donor Management* system to handle the online side of things, including email appeals, online donations, and website content management. The fundraising team is also considering adding a *Broadcast Texting* component to the mix to capitalize on the number of people who attend the River Days event each year, but are not members. "Everyone has a phone and it's so easy to text. If we have a short code up at River Days, they just punch in the number and then we have a way of communicating with them all year," said Susan Berg, Communications Director.

Major donors are a big part of the organization's fundraising. To reach them, the organization first created a core "major donor" team made up of development staff, board members, and a few committed volunteers. Since team members work across the state, the organization facilitates their work with collaboration tools. First, *Email and Calendar* software, which is set up in on a separate server, lets them trade email and share their schedules with one another. An *Email Discussion List* and regularly scheduled *Online Conferences* allow them to trade strategies and keep in touch. *Collaborative Document* tools let them work together on shared materials, and an *Intranet* gives them access to shared organizational resources.

*Project Management* software is also important to the Council, especially in the months leading up to River Days. Each team tracks its progress on various tasks and manages each member's responsibilities and deadlines, which makes working from different locations nearly seamless.

The Council has recently identified a need to better manage staff members, payroll, and hiring, so they're considering an **HR and Office Management** system to streamline that process and free up the time spent on human resources. Up until now, Drury has been managing employee data using Excel spreadsheets, which worked well when the organization was much smaller, but as the number of employees has grown, so have its needs.

Like the staff, the board of directors—key to the organization's continued success—is made up of busy people who are rarely in the same location. **Board Support** software lets members communicate with each other and with the executive director, and makes sure they have the right materials to review and comment on as well as a regularly updated calendar of board and organization events.

The Council relies on **Social Media** to create buzz around River Days and with the help of **Online Listening** tools communications staff are able to find and share photos and comments made by event participants. The Council is also trying **Online Advertising** on websites and blogs that cover topics related to their mission, especially in the months leading up to River Days. In addition, the organization tried an online pledge encouraging people to commit to protecting Wisconsin's waterways. The experiment went well—thousands of people signed the pledge.

Overall, the Council has found that the best software approaches help them to harness the passion and energy of the people of Wisconsin. "We have great people here," Drury said. "When we need their help, they always step up."

*\* All case studies in this guide are fictitious; they are used to explore how technology can help to address common challenges facing nonprofits.*

# Connecting With Students to Accelerate Achievement

## By The Book*
*Eight staff members/$860,000 annual budget*

Jason Bennett, Executive Director of By the Book, almost didn't go to college. "My mom said we didn't have the money and when I looked around I didn't see a way," he said. He was a B student, but he stopped going to class and stayed home playing video games. "Ms. Burns, my guidance counselor, came to my house one day and said, 'I'm going to make it my personal mission to graduate you and send you off to college.' I'd never had anyone back me like that before. It made everything possible."

Bennett became an A student and got into college. Then he graduated with honors. Ms. Burns was at his graduation and asked him: "What are you going to do next?"

"I'd been thinking about it for a while," Bennett thought, "but it took telling her to make it real. I said, 'I'm going to do what you did. I'm going to get kids like me to college.'"

Six years later, the nonprofit he founded to accomplish that goal, By the Book, is graduating its first class. To help celebrate this achievement, the Scholar Foundation just gave a $250,000 grant to increase the number of high school students it serves with its tutoring, college counseling, and life skills programs.

Next year By the Book will be in six high schools in the Minneapolis area. To track all of its donors, volunteer career advisors, students, and parents, Bennett decided to invest in a **Constituent Relationship Management** system (or CRM). Since there's a lot of overlap between these groups—for instance, former students often volunteer or donate, and some of the parents have seen more than one of their kids go through the program—Bennett wanted to be able to get a full view of specifically how each person is involved. This made a CRM system a better fit than either a **Donor Management** system, which couldn't easily track their program participants, or a **Case Management** system, which was too narrowly focused. He also considered a **Specialized Constituent Management System** such as a student information system, but didn't find the systems she demoed flexible enough for his needs. The CRM system also provides basic **Broadcast Email** and **Event Registration** functionality, integrating these online components while eliminating the need for additional software packages.

By the Book also has a separate **Online Donation** tool, which it chose because the costs were lower than the other tools it considered. However, Eric Santana, By the Book's development director, has to export data from that system and then import it into the CRM system, which is time consuming and sometimes leads to errors. Bennett recently met with a consultant to figure out if any **Online Donation** tools integrate well with By the Book's CRM system.

"We still have a few issues to work out," Santana said. "But we're getting there."

The biggest strengths of the CRM system Bennett chose are the built-in communications and fundraising functionality. It's easy to send an email newsletter to subscribers each month, create labels to mail paper newsletters, and support quarterly appeals. An **Email Discussion List** that lets board members, major donors, and alumni discuss ideas also felt important. Bennett worked with Lisa Tam, the organization's communications director and "accidental techie," to set it up and develop community guidelines. The list has turned out to be a bigger-than-expected success in understanding what's important to those groups, and engaging them with opportunities to help incoming students.

As it's important to the organization to proactively understand what its students need, Tam installed a **Website Analytics** package to get a sense of what online resources were getting used the most. She's also very aware of how active students are on social media and uses **Online Listening** to monitor what people are saying about By the Book. Every quarter, she also circulates an **Online Survey**, both via email and **Broadcast Text Message**, to more proactively check in with staff and students. Bennett, Tam, and Santana have discussed using a **Dashboard** to make it easy to see all of these different metrics in one view, but they're not sure if it would be worth the time to set it up.

A lot of By the Book's programs are based around specific events—for instance, a big SAT kickoff to encourage students to sign up to take the test. The organization also does a lot of promotion to make sure every student is aware of the opportunities that are available at their school. A **Content Management System** makes it easy to post events and information on the organization's web page, but Bennett and other staff post on social networking sites too—particularly **Facebook** and Snapchat, which are widely used by students. Using **Event Registration** software, students can RSVP to events and get on a mailing list for event updates. This streamlines the process somewhat, but it's always a little hit-or-miss to try to figure out who's coming.

Each event involves a lot of coordination, and staff members rely on many different back office and collaboration tools to help keep everyone in sync. They couldn't function without **Email and Calendar** software, and an online tool provides **File Sharing** functionality to allow staff both in the office and at the schools to access documents from wherever they are. They also use a simple **Intranet or Portal** to post information, dates, and contact information for all team members.

Bennett invested in professional-caliber **Page Layout** software, and his team creates posters for schools and community centers, and infographics for funders and the school district. Not coincidentally, the software's also great for laying out newsletters. Photos are an important tool for By the Book to help show students working hard and succeeding. Tam convinced Bennett that they didn't need anything fancy in this area—in fact, free **Photo Editing** software meets all their needs for print or the web, and they can share whole libraries of them online using photo sharing websites and Instagram. The students love seeing pictures of themselves and each photo typically gets dozens of likes and comments on **Social Media**.

Early on, Bennett noticed how many of the students used their cell phones for everything—and how many of their parents used them to keep in touch. He worked with Tam to optimize their website to make it **Mobile-Friendly**, which was much easier than creating a separate mobile website. By the Book has also begun to use text messaging as a primary communications tool, not only broadcasting information to all of its students via text, but also automating individual reminders via an app in its CRM system and even texting individuals when a quick communication is needed. "We used to call them and they wouldn't get back to us. When we'd ask them why, they'd tell me, 'I don't have any more minutes!' Now we text them and get a response in seconds," Bennett said.

In the short term, Bennett wants the students in the program to participate in more events and not miss a single meeting or appointment. And someday he hopes they'll be volunteers or donors. No matter what, he's looking forward to each graduation. He said, "I shake each one's hand and I tell them, 'You opened the door wide. Now walk on through.'"

*All case studies in this guide are fictitious; they are used to explore how technology can help to address common challenges facing nonprofits.*

# Establishing a Community of Artists

## Rogue Artists*

*One staff/$180,000 annual budget*

Last year, a donor's estate gave Megan Fine, Executive Director of Southern Oregon Painters, a seed grant to create a nonprofit organization to support artists in Southern Oregon. The gift included a building in downtown Medford, several paintings by local artists, and a lump sum of cash. The donor's directions were to create a space that "helps artists continue their work and displays the vitality of the art community."

"When I heard about the grant I was just blown away," Fine said. "The community really needs this."

The seed grant will support Rogue's single employee, Fine, and the upkeep of the building for a few years—enough time for her to get everything up and running. It won't last beyond that, however, meaning that fundraising is a top priority. Rogue Artists also needs to reach out to its local community to let artists of all ages and abilities know about this new resource.

Fine started with the basics. She knew that she'd need **Office Software** and tools to support **Email and Calendaring**, as well as a way to track finances in a straightforward Accounting system. She uses free **Virus Protection and Firewall** software and inexpensive online **File Backup and Recovery** to ensure the safety of the organization's contacts and other materials.

"There was a lot to set up and I'm not a computer person, but I figured it out with a little help," Fine said.

With her software infrastructure in place, she began her fundraising process by creating lists of possible individual donors, foundations, and artists in Excel spreadsheets. Before long, she realized the spreadsheets would make it difficult to track multiple donations and the small, but important, number of artists who were also donors.

She invested in **Donor Management** software that can track all the people with whom she interacts and it's made a huge difference. She imported all the Excel spreadsheets she'd created with no trouble, and now can easily find people, enter gifts, and create lists of people she needs to call or mail. She considered a **Constituent Relationship Management** system, partly because she is managing studio space and classes at the center and was hoping to find an **All-in-One CRM** system, but decided she didn't actually need much additional functionality beyond what a donor management system could do and couldn't find a low-cost system met all her needs. Her calendaring software and spreadsheets could handle the logistics and the **Accounting** software handled the money.

Now she's looking at ways to keep connected with artists and the Southern Oregon community. "The community is what keeps us going," she said. "They're my top priority." She's considering creating an **Email Discussion List** to help artists keep in touch with each other—and with the organization—about issues of mutual importance. She also knows the value of a good **Website**, so she hired Matt, a consultant, to build the site. She chose Matt primarily for his experience with free **Content Management Systems**. He showed her how to use the software to update the site's text and images so that once the site is built she can keep it fresh.

Fine is working toward adding an online "virtual tour" of the space and a digital gallery to show off new work by artists who rent studio space. **Photo Editing** or **Multimedia Editing** software are now easy enough to use that she can create and edit images herself, saving her from hiring a photographer or graphic designer. She's also discovered that many of the artists she works with also willing to donate their time and expertise. She's considering **Crowdsourcing** an upcoming advertising campaign and has already set up **Collaborative Documents** to collect and manage the work that comes in.

She also created *Facebook* and Pinterest pages for the organization, and tries to spend at least two hours a week posting updates. However, since her constituent base is not likely to seek out a more steady social media presence from the organization, she's relegated such actions to a lesser role. Because so many of the young artists in the community have *LinkedIn* profiles and part of her mission is to professionalize many of the talented, but struggling, artists in Southern Oregon, she created a discussion group through the site that links them into an informal community where they can network and start conversations. She also uses *Online Conferencing* to "meet" with artists as she works with them to plan upcoming exhibits.

What's next on the software front? Fine's considering the best way to create a newsletter. Initially she wanted a printed newsletter, and looked into *Page Layout* software to help create it, but now she's leaning toward an email newsletter. If she had a solid *Broadcast Email* package she could use it to send not only eNewsletters, but email fundraising appeals, as well, which can help her meet her organization's mission of being completely self funded within three years. *Online Donation* tools would help her handle the resulting gifts and pledges. She's also prioritizing online foundation grant research to find other grant opportunities to help keep her organization in existence once the seed grant runs out.

She's confident that with the right software she'll be able to successfully run the entire organization and serve her audience well, despite being the only staff member. On the wall next to her desk she posted the letter the donor wrote to her. It says, "Help our area make something beautiful." She looks at it every day and it's become her personal mission.

*\* All case studies in this guide are fictitious; they are used to explore how technology can help to address common challenges facing nonprofits.*

THE FIELD GUIDE TO SOFTWARE

*Every nonprofit, regardless of its mission, needs a set of software tools to help with day-to-day work. From Office Software to Email and Calendar to Virus Protection, these tools can help you be more efficient and effective.*

*All of the software names highlighted within the text are covered in more detail in this guide. Most of them are included in this section and follow immediately after the descriptions.*

BACK OFFICE AND PRODUCTIVITY

# Managing Productivity

In addition to the Accounting, Office Software, Email and Calendar, Virus Protection, Firewalls, and File Backup and Recovery systems covered in the Every Organization Needs section, software designed to help with the day-to-day activities of any organization can make your work more efficient and hassle-free.

## Strongly Consider...

If you want to take users' credit card information for memberships, donations, purchases, or other transactions, you'll find a number of affordable **Credit Card Processing** systems that can make such tasks painless and secure.

## Keeping Ahead of the Curve...

Passwords are often the biggest vulnerability at a nonprofit. People often write their passwords on post-its or create simple passwords because they have so many passwords to remember and don't want to get locked out. **Password Management** software can make it easy for them to keep track of passwords while keeping them secure.

Larger organizations can create a considerable volume of documentation. It's a lot easier to create than to store and manage this output, especially when you go back to find something later. **Document Management Systems** can help you keep the mountains of materials under control and within reach.

On the Cutting Edge...

Larger organizations may want to explore dedicated **HR and Office Management** software to keep track of hiring, benefits, and many other tasks related to your most important resource—your own staff.

# Accounting Systems

If you're managing more than a few thousand dollars at a time, you almost certainly need an accounting system. Even if you have help with accounting—maybe bookkeeping, payroll, or someone to help out at tax time—you still need a solid system to track and manage revenue, expenses, payments, and other finances.

Conversely, even if you have a good accounting system, you still need some knowledge of accounting. If you lack that knowledge, consider hiring a bookkeeper to set up the software so that it does what your organization needs it to.

Lower-end systems starting at a couple hundred dollars per year will work for most organizations with one or two accounting users and budgets of up to $2 million. QuickBooks is popular and relatively easy to get up and running. Its primary products are online, Cloud-based services that can be accessed on a desktop or a mobile device. Sage offers multiple products for small nonprofits: Sage One is online accounting software that allows up to two users and can be accessed on both Apple and Microsoft devices; Sage 50c is an installed system that allows up to 40 users and can also be accessed online. Other systems, such as FUND E-Z, are designed specifically for nonprofits and provide features that are of interest to nonprofits that need to track money in and out of complex funds. However, FUND E-Z is much more expensive than the other options, with pricing starting at $2,000 for one user.

Organizations with more than four or five users, budgets over $2 million, or a large number of funders might consider more complex systems. At this level, the cost of implementation can sometimes exceed that of the software, although you often can choose to host the data yourself or host it in the Cloud. Nonprofit-friendly systems in this realm include Abila MIP Fund Accounting and The Financial Edge by Blackbaud. QuickBooks also offers a more robust version called QuickBooks Enterprise, which starts at about $1,000.

Not all organizations need systems created specifically for nonprofits. General business accounting packages, such as Microsoft Dynamics, NetSuite ERP, and Sage 100c, tend to be more widely used, which makes it easier to find help with setup, support, and bookkeeping. They can also track inventory, billable hours, or other relevant data, and offer support not provided by most nonprofit-specific packages. And while some systems do not explicitly cater to nonprofits, some offer nonprofit pricing.

Accounting is a complex area, and there are a lot of high-end systems available. If you manage hundreds of millions of dollars, more than a hundred staff, or a complex, multi-unit national or international structure, seek expert advice on an enterprise system.

## Don't Forget Payroll

Nonprofits with multiple full-time staff members (rather than volunteers) are going to need some way to manage payroll, time, and attendance. Small organizations might look to Intuit's Online Payroll (formerly PayCycle), while mid-sized organizations might look to ADP (Automatic Data Processing) and time-tracking tools like Kronos. The smallest nonprofits may be able to manage these processes through Excel or Google Spreadsheets.

# Credit Card Processing

Want to take payments via credit cards, either online or in person? It's neither difficult nor expensive to do.

First, you'll need to choose a payment processor. There's a good chance your bank offers processing services, but shop around—some banks charge many complex fees. Popular credit card processors include Authorize.net, SafeSave Payments, and Dharma. You'll also need to set up a merchant account, which is where the money goes when you receive payment from a credit card and where your fees are taken out.

Online payment processors, including Click & Pledge, Greater Giving, and GiftTool, allow you to process credit card payments (such as online donations, event registration, or item purchases) over the internet, typically for a $20-$40 monthly fee, a small percent of the transaction, or a per-transaction fee. These vendors typically provide an interface optimized for your constituents to submit payments on their own, but most of these interfaces also work as input screens for staff to process payments.

Alternatively, many **Donor Management** software packages—such as Little Green Light, DonorPerfect, NeonCRM, and The Raiser's Edge—let you process donations and other payments directly from that software. This convenient option lets organizations process a high volume of a single type of payment, and saves time-consuming double-entries.

If you need to accept payments at an in-person event, you'll need card reading equipment—often called "swipe terminals." Your swipe terminal should include an EMV chip reader and your processor should be able to handle EMV chip cards—the new standard for credit and debit cards with a tiny electronic chip embedded in them. This is an important recent development that puts the liability for fraudulent charges on merchants (you). If you swipe the card's magnetic strip when it has an EMV chip, and the card was cloned or stolen, you will not receive payment for those charges.

Also, you should purchase your terminal rather than renting it, since it is generally low in cost. In fact, it's a big red flag if your processor wants to lease the equipment to you because the contracts are notoriously difficult to get out of and the total cost is often many times the real value. If you only need a terminal for a one-time or occasional event, Greater Giving's Auctionpay and similar sites rent terminals with a focus on nonprofit events.

Many nonprofits now process credit cards at in-person events using smartphones or tablets. This ultra-portable method uses a mobile device with a data plan or Wi-Fi to process transactions by letting you either manually enter card numbers or, with the addition of inexpensive hardware, read cards directly. Many vendors have emerged in recent years, including PayAnywhere, Square, PayPal Here, ROAMpay, and Intuit GoPayment. Additionally, many traditional processors and fundraising software also offer mobile apps that allow you to accept payments. Many of these processors provide free swipe equipment, but not all mobile processors provide chip reading equipment.

If you need to take credit cards onsite and a mobile or terminal payment processor isn't practical, you can use an imprinter—those small machines that swipe and carbon copy the credit card so you can charge it later. Just make sure you have policies in place to protect payment data from loss or theft, including securely storing and destroying slips when they are no longer needed.

If you want to take credit cards at a permanent physical location such as a gift shop, registration desk, or cashier station, you'll need more hardware. Point of Sale software, such as Worldpay, CAM Commerce Solutions, or Keystroke POS, can integrate it all and help you manage actual inventory. Square and PayPal also offer point of sale equipment and functionality.

# Document Management Systems

Sometimes your day comes down to finding a piece of paper with one key bit of information on it. Document management systems promise to streamline that process, whether it entails walking to a file cabinet and fishing out the necessary folder or rooting around piles on your desk.

At its simplest, document management can be a shared drive with a clear file-labeling protocol created by your organization or an intranet with essential documents, from core messaging to reimbursement forms (see *Intranets and Portals* for more). But software dedicated to the purpose provides a virtual file cabinet in place of storing information on paper. It can reduce time spent finding documents and help you maintain them securely by controlling access with permissions, similar to providing necessary staff the keys to a physical file cabinet. (It also takes up less physical space than paper files.)

The first challenge of a paperless system is getting everything into electronic format. If you have a multi-function copier, the same company that sells your hardware likely will bundle a document management software solution for an additional fee. Konica-Minolta offers proprietary software that includes Optical Character Recognition (OCR) software and Xerox DocuShare comes with an app to access documents via iOS or Android smartphones.

These tools can scan a document and recognize key text such as an invoice number or client name, and file the electronic copy automatically. For documents created electronically, document management software lets users print, save, or file from Microsoft Office or other software directly to the organization's virtual file cabinet.

Retrieval is where greatest time savings comes in, providing documents are saved in a way that users can find them easily. System-generated metadata, such as creation date, author, and type, as well as keyword and full-text search are other document-management retrieval methods. Since keywords tend to reflect the unique mission and operations of each organization, document management systems typically involve an element of customization at the outset.

Customization is one key factor in driving cost of these systems. One way to avoid the cost of customization is to rely on a product your organization may already own, such as Microsoft Sharepoint, which incorporates basic categorization through its Document Center feature.

eFileCabinet starts pricing with a $500 implementation fee to get the software up and running for three individuals, and charges $600 per user, per year. Fees go up for increased storage, more users, and increased levels of complexity, such as API access. Firms like OpenText and Alfresco have evolved to handle types of content beyond documents—videos, for example. Costs may be higher, but these more elaborate products include robust support and education. For example, Laserfiche (whose services also start at $600 per user, per year) provides extensive training and education.

Two other factors to consider in pricing a document management system are how many people on your team will need access to your virtual file cabinet and how many concurrent users you might have.

# Email and Calendar

Linking your calendar to your email lets you schedule meetings and invite others to them, respond to invitations from other people, and share your work schedule and availability with coworkers or select people outside your organization. The most common tools for email and calendaring are the flagships of Microsoft and Google: Outlook, the email and calendar app at the heart of Microsoft Office; and Gmail with Google Calendar, part of the recently renamed G Suite (known as Google Apps prior to September 2016).

The two offer similar functionality by integrating your list of contacts and to-do lists, and both offer versions for solo users and entire organizations. Outlook also includes a note-taking application.

G Suite is free to download for nonprofits, but if you want your own domain name—for example, name@yourorganization.org— you'll need to pay to acquire and maintain a web domain host.

A comparable Cloud-based version of Outlook is also free via TechSoup as part of the Office 365 Essentials service, which includes online-only versions of Microsoft Office and other tools. To run a full-figured desktop version of Outlook, nonprofit pricing begins at $2 per user, per month. Additional features, such as HIPAA and other compliance tools, better security, and more administrative control over user accounts, are also available for additional cost.

Standalone "perpetual-license" versions of Microsoft Outlook for either PC or Mac are also available via TechSoup. These are suitable for organizations based in a single physical office with the ability to run a client-server arrangement.

In addition to the standard-bearers, there are scores of other email and calendar options—some free, others available to download for a fee. A few of the better-known tools include Mailbird, Mozilla Thunderbird (from the same nonprofit that makes the Firefox browser), Apple Mail and iCal, and Airmail.

# The Name Game

You'll need to register a domain to use any organizational email tool. Using the same domain for your organizational email accounts (george@idealware.org) and website (www.idealware.org) is inexpensive, looks more professional, and is easier for people to remember than using free email domains like gmail.com, yahoo.com, or hotmail.com.

Use an online domain registrar to see if a particular domain is available—NetworkSolutions.com, Register.com, NameCheap.com, easyDNS.com, and Gandi.net are some common ones—and lease the rights to the domain for $5 to $20 per year. Someone with a bit of technical savvy can use the tools provided by your registrar to point your new domain to your email software package. It's worth noting that some registrars are better than others at protecting clients from false takedown notices. If your work is controversial, read up on the registrar's policies regarding domain security. Do they shut down first and ask questions later when a business or government takedown request comes in?

You might also explore domain registrars with built-in, easy setup options with third-party software. Google Domains is a familiar choice and can integrate with Google Apps. Others, such as iwantmyname.com, have built-in setup options with a variety of applications, such as customer service and broadcast email systems.

# File Backup and Recovery

Backing up files is critical in case of damage, loss, or theft, but file backup is a process more than it is a single piece of software. A good backup process contains three elements: ease of use or automation, offsite or Cloud storage to prevent your backups from being destroyed along with your primary data, and the ability to retrieve those files and restore your data.

For individual users, newer versions of both Windows and Mac OS include file backup and recovery tools that can be used with either external hard drives or with Cloud-based storage. There are also a number of third-party Cloud-based backup and recovery options available for individuals. Unlike simple storage options, these systems help automate the process so that it is not dependent upon users remembering to make copies of files. They also include tools to create disk images of an entire computer—including applications, data, and personal settings—to make it easier to get back up and running after a loss.

Some tools, such as iDrive, offer free backup services for small amounts of data. In addition to more capacity, paid tools offer more robust and sophisticated features. Versioning—the ability to roll back to previous versions of specific files—is a mid-level service included in some paid solutions, including Carbonite. MozyPro (now part of Dell), Crashplan, and JungleDisk offer software and storage to make both local and Cloud-based backup possible.

To back up an entire office of computers or a network, a common solution is what's called a Network Attached Storage, or NAS, device—essentially multiple hard drives in a single cabinet or rack that stores redundant copies of data. If your staff works in the field using laptops or mobile devices, they can also back up to a NAS system remotely over the internet.

A local backup is usually faster than Cloud-based storage when recovering an entire file system, but is not sufficient to protect against fire or other damage to the office. When selecting backup and recovery software for an office or network, consider whether it will allow for backup to both a local NAS device and to Cloud-based storage, which is a best-of-both-worlds solution.

## What Should You be Backing Up?

The simple answer? Everything. Or at least, everything that's critical, valuable, irreplaceable, or important to your organization. Consider each of your organization's processes separately and account for each by identifying and backing up its data. And remember—backing up file servers won't do any good if individual staff members are storing their valuable data on local hard drives. Make sure networked computers all have access to file storage, and that everyone on staff is using it.

# Firewalls

A Firewall is not exactly a wall—it's more of a gatekeeper that decides what information can pass based on predefined rules. You can install a firewall at any point where networks or devices meet—including routers, switches, and computers—to guard against intrusions from hackers and malicious software.

A good Firewall includes predefined settings that let common traffic pass unimpeded—otherwise, users need to approve every transaction, which can be both overwhelming and annoying. It should also enable you to whitelist or blacklist specific IP addresses or hosts, approving or denying access on a permanent basis.

A Firewall that runs in stealth mode makes the computers it protects less visible to hackers and similar threats. More advanced Firewalls can inspect all information from the internet for evidence of intrusion into your systems or attempts to disrupt access to them, and filter out files that are likely malicious.

Network-based Firewall appliances protect your entire organizational network by inspecting traffic from the internet and screening out malicious activity or information before it reaches your computers. A Firewall appliance may be as simple as a home/office router from TP-Link, Linksys, or Netgear, or as sophisticated as the enterprise products from Cisco or SonicWALL. Prices vary widely depending on features and functionality. A low-end Firewall appliance can be bought for under $200, while a high-performance device can run several thousand dollars.

However, most threats are more sophisticated than what the classic Firewall can detect. New Firewalls, often called next generation Firewalls (NGFWs), are able to block hacks and malicious software at the application level rather than just the port and protocol levels—this means you can set controls at the application and device level to account for a diverse array of threats. They can also more thoroughly inspect the information that is passing through (called "packets"), filter websites based on their source and content, manage bandwidth, and receive real-time updates. Most Firewalls also allow you to change settings remotely using a mobile or web app. Cisco's Meraki and FirePOWER, Dell's SonicWALL, and NGFWs from Palo Alto, Fortinet, and Check Point all earn high praise from experts.

In addition to NGFWs, some organizations use Secure Web Gateways (SWGs) to enforce security policies on web surfing devices. Blue Coat and Zscaler are two top products.

Firewall software installed on individual computers can provide machine-level protection that a Firewall appliance can't. For example, a laptop without its own Firewall might be safe while connected to your protected organizational network, but will be vulnerable when connected to an unfamiliar wireless hotspot.

# HR and Office Management

Human Resources Management Systems and Human Resources Information Systems—commonly abbreviated as HRMS or HRIS—can reduce administrative time and improve efficiency by helping your organization track and organize its human resources data.

If you don't have a dedicated HR department (or even a staffer whose sole duty is to manage human resources), you're not alone. Many nonprofits don't. They rely on employees with responsibilities in such other areas as administration, finance, or operations to take on HR as well. If this is the case at your organization, user-friendly HR software that doesn't require a lot of upkeep can become a time-saver for the employee who uses it.

How do you know if you need an HRMS? That depends on how complex your rules and benefits are. Generally, if your organization has 80 or more staff members, you should consider implementing software to help your HR manager respond to needs in a timely manner. If they have no HR staff, even smaller organizations can benefit from an HRMS because it can save a lot of time that can be used for other projects. For organizations with fewer than 25 people, however, these systems are likely to be unnecessary.

At its most basic, HR management requires a place to store all employee information. If all you're tracking is basic information, you can use Excel or Google spreadsheets. It's possible to create a more powerful solution using Microsoft Access, OpenOffice Base, or FileMaker databases to manage employee demographic, contact, and compensation information. You may also find it possible to expand your existing payroll software. Solutions such as Intuit's Payroll Services or Sage's Payroll Full Service integrate with Intuit's QuickBooks accounting software and can provide functionality to track benefits, employee demographic information, and compensation history, and can serve as a useful repository for employee data in lieu of a dedicated HR system.

If your organization is more than a few dozen employees, consider implementing a more-advanced solution, such as ADP HR Management Software or Sage HRMS, which can handle more complex timekeeping issues. For the largest organizations, you may need an enterprise-level solution, such as Kronos Workforce Ready, Ascentis, Ceridian, or BambooHR.

You may also consider a smaller, specialized solution targeted at managing only a specific subset of HR. These include Newton, Workable, and iCIMS for hiring and recruitment, and Shiftboard or When I Work for scheduling. These tools can be used in conjunction with one of the smaller, less feature-rich solutions in order to meet a specific HR management need without investing in one of the more expensive enterprise solutions.

# Office Software

Office software—also called productivity tools or productivity suites—includes the programs we use to write, track numbers and make lists, create presentations, and a constellation of other related day-to-day tasks. Microsoft Office remains the best known—Office 2016 includes Excel, Outlook, PowerPoint, Publisher, Word, and OneNote, a note-taking application that allows various types of content to be shared among team members. The slightly more-expensive Professional Plus version adds Access, InfoPath form builder, and the Skype for Business team-communication tool.

Recently Microsoft has begun encouraging users to move away from the "perpetual license" model and toward Office 365, its subscription model of software licensing. Hosted, or Cloud-based, editions of Office tools are available at five levels of service with other features (including more software or advanced teleconferencing and voicemail support) available for additional cost.

How to choose between the desktop and subscription versions? Each desktop license allows users to install the software on two devices, but Office 365 can be installed on multiple devices—including tablets and smartphones—and adds collaboration tools and online storage. The desktop version also includes limited software updates, while Office 365 users always have access to the latest version. If you have a single location and staff tend to restrict their work to the office, the desktop version may be the better choice. Office 365 is better for staff on the go, or who work on many devices.

In many ways, Office 365 is Microsoft's efforts to compete with Google's hosted suite of productivity tools, of which the major rival is G Suite (formerly Google Apps). G Suite includes Docs, Sheets, Slides, and Forms—direct competitors to Office's best-known tools—as well as the simple webpage creator Sites, Calendar, Hangouts videoconference and chat tool, Google + social media platform, and Google Vault archiving service. While the two suites offer different strengths and weaknesses, one area in which they're on par is document collaboration and reviewing. Both platforms support multiple authors in a document at the same time, and both provide some ability to track and review changes and comment on and chat while editing drafts.

Other alternatives exist, including the open source suites LibreOffice and Apache OpenOffice, which offer similar functionality at no cost, though neither offers a Cloud option. Mac users have access to the iWork suite bundled with their OS.

File compatibility among platforms varies. Most files can be opened and edited by suites other than the one in which they were created, but formatting and other features might be lost.

# Password Management

Passwords continue to be a major threat to an organization's data. Weak passwords (the most popular password of 2015 was "123456"), writing passwords down and leaving them at your desk, writing passwords in emails, or reusing passwords are bad habits that carry a lot of risk.

Why do people adopt these bad habits? According to an Intel Security survey, the average person has 27 accounts that require a password. It's unreasonable to expect someone to remember 27 complex, unique passwords.

Password management software has emerged to help organizations and their staffers deal with the overwhelming number of passwords they need to manage. A typical password manager acts as an auto-fill function in your browser that is activated when you enter the password for the management service, essentially allowing you to use a single password to access all of your passwords.

You can set up your password manager to require you to log in once or every time you need a password. For users not comfortable with the auto-fill function, another option is to open a secure application that stores all of your passwords and allows you to view the one you need. A good password manager is encrypted, uses Secure Sockets Layer, or SSL—a standard security technology for establishing an encrypted link between websites and browsers—on the web, enables multifactor authentication, and will audit your passwords for strength and duplication.

While many services offer free accounts for individual users, nonprofits that want to provide the service for multiple users would need to pay for an enterprise account. Enterprise accounts typically allow you to set permissions across your organization, log in to installed apps, fast-track new account setup, securely share passwords with specific people, and sync passwords to multiple devices. Enterprise accounts with LastPass, Dashlane, and Keeper cost $20 to $25 per user, per year.

# Virus Protection

Antivirus software serves two roles: to stop malicious software—known as viruses or "malware"—from reaching your computers; and to "disinfect" computers of malware that has already been installed. While antivirus software is typically installed on individual laptops and computers, there are also solutions designed to protect email servers and file servers.

Traditionally antivirus tools have been PC-focused, but most antivirus solutions now also offer versions for Mac. Norton, Kaspersky, and Webroot are popular paid desktop antivirus solutions, but many nonprofits are turning to the free versions of tools such as AVG, Panda, and Avast, which offer competitive basic functions. Keep in mind that paid antivirus software often offers centralized management, which can be valuable for organizations with more than a few computers to protect.

Antivirus applications should offer real-time protection by monitoring machines around the clock, scanning incoming and outgoing emails and attachments for viruses and quarantining any they find. Make sure the antivirus solution you choose will integrate with your email client, if it's not web-based.

You should also be able to schedule periodic updates to let the software download new virus definitions to protect against all current-known threats. Automatic updates are particularly helpful for organizations with multiple employees who might not all remember to run updates manually.

While antivirus solutions for regular file servers are typically the same as those for desktops and laptops, email server antivirus usually requires an additional cost module that scans and cleans messages before making them available to users. The goal is to keep danger as far away from your users and systems as possible. Most of the major antivirus vendors offer an installed version for email servers.

Many organizations now use Cloud-based email antivirus solutions from vendors such as Panda and Symantec to detect and clean any infected email messages before they reach the network. Some hosted email solutions, such as Google Apps, will include this automatically.

Emerging so-called "next generation antivirus" software is creating buzz. At least two of these products—Code Black and SentinelOne—have been certified to replace traditional antivirus software. These systems use whitelisting controls, endpoint security to validate users, and pattern recognition to detect unusual activity. Pricing varies by the size of your organization, but a midsized nonprofit can expect to pay hundreds of dollars per year.

Anyone who wants to compare antivirus solutions can go to av-comparables.org, a site that runs independent tests of many of the top antivirus software on the market. You can also do your own sleuthing at http://scanurl.net/, where you can check to see whether a link is safe before clicking it.

## Don't Forget the Human Element

Virus Protection doesn't stop with software—your staff members have to use their common sense by not opening suspicious emails. It's also important to change passwords regularly and to create more secure passwords (stay away from "password" and "1234"). Read our sections on *Password Management* and *IT Security Policies* for more information on how to keep your data safe from hackers and thieves.

*Do you know how well your programs are performing or what people are saying about your nonprofit online? Data analysis practices and tools can help you evaluate your programs, understand how and why people come to you, listen in on what people are saying about you, and discover opportunities to make a bigger impact.*

ANALYTICS

# Analyzing Your Organization's Data

Your organization probably collects a lot of data—and if it doesn't, it should. Maybe you record case details in a *Legal Case Management Software*, collect *Survey Data* from clients during an intake or after a case is over, maintain a *Website*, or post to *Social Media*. All of these systems collect useful data. You can also find public data created by other nonprofits or government agencies that can inform you about your client base or the community you serve.

There's a wealth of data available to you. The question is: How do you use it to improve operations and make your programs more effective?

## Strongly Consider...

Whether you needs include *Analyzing Paper Data* or working with data that's being collected by a database system, you need to view it in an analyzable format. Your case management system likely provides reporting features that enable you to see the data right there within your system. However, if you data source does not allow you to see charts and graphs or manipulate the data in useful ways, you may need to pull the data into a spreadsheet.

If you're dealing with multiple data sets, then you'll need to match up the data so that you're not seeing duplicate records. For example, if many of the people in your case management system are also in your broadcast email system, you'll need to combine the information that belongs to users that show up in both systems, otherwise you'll have a dataset full of duplicates and incomplete data. There are automated ways you can do this, but you might need expert help to pull it off.

Once you have a solid list to analyze, you'll need to clean your data. This means making sure that, overall, what you're viewing is accurate. To clean your data, check whether fields or whole records are missing. If so, try to find out whether there's a systematic reason why. For example, if one office is not collecting client data, or one person keeps forgetting to record employment status, you will have a big hole in your data that will lead to skewed analysis. Make sure notation is consistent—are staff entering state initials or spelling out the state name? It might not seem important, but inconsistent entries will make analysis very difficult. Finally, scan the fields for anything that looks unusual. For example, if you work with children, but you see a birth year listed as 1954, then something is not right and may require some investigation.

Once you have a reliable body of data, you should ask yourself whether you have enough data and whether there's any collection bias in it. If you feel you have a representative sample of the people you want to know about, then your data is ready for analysis.

Analyzing quantitative data can be as straightforward or as complex as you want it to be. Some people rely on looking over graphs to help them see trends or outliers. Others sort the data or create new data using spreadsheet formulas such as SUM, AVERAGE, and COUNT IF. Pivot tables are a little tricky to set up, but are powerful tools for comparing aggregate data from multiple perspectives. (Not familiar with these spreadsheet features? Consider online training from Ann K. Emery, Lynda. com, NonprofitReady, or other organizations that provide training in this area.)

You can also analyze qualitative data such as comments, case notes, and stories. One way is to summarize based on your observations and interpretations. However, a more rigorous approach would be to break comments or quotes into chunks and code each chunk according to topic or theme. It's helpful to use a spreadsheet where you enter a statement in one column and codes in other columns. For example, you can code every mention of violence with a "Violence" tag in the coding column. Then, after you've gone through all of the statements, you can count up how many times that code appeared. From there, you might divide the number of times the code appeared with the total number of statements. That will give you a percentage of the number of times "Violence" was mentioned.

These simple techniques and others can help you turn a mess of numbers and comments into meaningful information.

## Keeping Ahead of the Curve...

If showing your outcomes to funders and supporters is a high priority, you may need to go further than the techniques mentioned above. Some organizations have established program evaluation teams that develop a series of metrics tied to the organization's mission and theory of change.

Consultants with data analysis expertise can help you take a closer look at your programs or operations and show you what's working and what isn't. If you are a large organization with an emphasis on data, you may even need to hire staff to run your data program. These data experts will likely use **Custom Reporting Tools** and implement **Dashboards** that automatically update when new data enters the system. They may also draw on other data sources, including open data and **GIS Systems** to help you contextualize your work against what's happening outside your organization.

Whatever approach you take, remember that every nonprofit has data and anyone can use that information to make their organization more efficient and effective. It takes a little time and attention, but the payoff is often significant.

# Analyzing Paper Data

Organizations that are surveying constituents or gathering client-reported data might find it impractical or too expensive to use software to do so, especially when working with populations that are unfamiliar with or unwilling to use *Survey Data*. However, the quality and affordability of mobile devices and the increasing number of apps that are compatible with other software tools make it feasible to collect data in person through a guided process and record it digitally in one step. Most data experts recommend that you explore a mobile option before resorting to paper data collection because the time and expense of managing paper is often equal to or greater than the cost of a technology-based solution.

However, if paper data collection is the only way, there are tools that can help you manage the mountain of paper and convert your data into a digital format.

One solution is Optical Character Recognition (OCR) software. OCR is a process by which handwritten or printed text can be digitized into a computer using an external scanner. That image is then converted to machine-readable text that can be searched, analyzed, edited, and/or imported into your database. OCR programs can be fallible, especially when dealing with messy handwriting or nonstandard fonts, but their accuracy has improved vastly in recent years. Staff members will likely need to take the time to check over the scans and correct them manually.

New scanners often come with basic OCR software. If you're looking to scan a large number of documents at once, look into a dedicated document scanner with a feeder that can allow you to scan whole stacks of paper at rates of up to 150 pages per minute. The staff time this can save you might be worth the investment.

If you're on a tight budget and have a newer scanner, consider freeware OCR software, such as OCRFeeder, FreeOCR, Tesseract GUI, or TextRipper. However, many of these tools require a strong technology background. Additionally, if you use Microsoft Office, you may already have Microsoft OneNote installed, which includes OCR capabilities.

Once your paper data has been scanned and digitized, you'll need to analyze it. Sociologists, anthropologists, and other academics that use Qualitative Narrative Analysis (QNA) to conduct their research often rely on tools called Computer Assisted Qualitative Data Analysis (CAQDAS) to discern patterns or trends from many pages of text. QNA software automates the coding process through user-defined "story grammar" that assigns categories and values to narrative content. Proprietary software can be very expensive, and could be of limited use to a nonprofit that only needs to analyze a few hundred interviews. Major players on the market include NVivo, ATLAS.ti, and QDA Miner from Provalis, which also offers a free and fairly robust version of the software in QDA Miner Lite.

Open source CAQDAS tools are often developed and supported by universities. Coding Analysis Toolkit, a web-based system developed by the University of Pittsburgh and run by Texifter, is a free and user-friendly option if you have basic needs. Qiqqa is described as "reference management software" and helps manage and analyze information stored in PDF documents—and also includes a built-in OCR process. If you're conversant in the programming language R, and want to use it to analyze your data, RQDA is an R package designed for CAQDAS. Like R itself, it's free and powerful, but you'll need to understand the coding language to use it.

# Custom Reporting Tools

While most Case Management or Constituent Relationship Management (CRM) systems will provide a range of flexible and useful reports, nonprofits looking to expand their report repertoire may want to consider designing their own reports using dedicated custom reporting tools. These reports can help with more complex evaluation efforts compared to built-in reports.

One solution would be to use a custom reporting tool to augment your existing capabilities. Tools such as Microsoft Access and Filemaker Pro are inexpensive, flexible ways to build your own database and customize your reports. While these databases may be easier to use than other custom reporting tools, and can easily import and export data to your constituent database, you'll need to create and maintain your own documentation so future staff members will be able to understand how the system works.

Google is currently beta testing a dashboard and reporting tool called Google Data Studio. It claims to offer "pre-built data connectors" that will make it easier to feed external data into the system. And, as you would expect, it is seamlessly integrated with every source of Google data. You can create up to five reports for free. To create more reports or to allow multiple report users, you can pay monthly, but pricing is not publicly available.

Crystal Reports from SAP is a widely-used custom reporting tool that provides the built-in reporting infrastructure for many software systems. However, while it's relatively inexpensive— about $500 for the license fee—the software is not the most user-friendly. Crystal is still good for building basic forms and handling registration data, but for complex data analysis, it might be more than most nonprofits need or want. Other similar reporting tools, which also offer what are often called "business intelligence" features, include Birst and Logi. JasperReports and BIRT offer two open source solutions. Two more tools that are good at bringing multiple data sources together include Looker and Zoho Reports.

You could also consider Microsoft's Reporting Services or Analysis Services, which come bundled with an SQL server.

# Dashboards

Nonprofits track all sorts of information, from financial data to event attendance, volunteer participation, supporter involvement, and more. A dashboard—sometimes called an executive dashboard—is simply a way to make it easier to understand and act on all this information by pulling it together in one place, with easy-to-understand visuals.

A good dashboard pulls together different, and sometimes disparate, metrics into a visually-appealing, easy-to-understand interface. Often it will show indicators that make it easy to see progress against a goal—for example, a "traffic light" icon with a green, yellow, or red light to show whether fundraising revenue is proceeding according to plan.

Deciding what metrics to track and where to find that data can be deceptively challenging, as can the logistics behind creating and displaying it as a dashboard. The simplest way to create a dashboard is to use one that already exists. Many **Constituent Relationship Management** and **Donor Management Systems** come with pre-programmed dashboards to track fundraising campaigns. For example, a flexible constituent database such as Salesforce might track enough of the metrics you want to look at to simply create a dashboard as a report.

Microsoft SharePoint, often used as an **Intranet and Portal** solution, provides dashboard features for those willing to invest the time to learn and implement the software.

A more straightforward approach is to use Excel, as you can easily paste in updated figures, summarize them on a highly formatted summary tab, and use charts and automatic color-coding to create something highly readable—though it's more difficult to make it beautiful. You could do something similar with Google Sheets, although it's not quite as powerful for creating complex calculations. If you're creating static reports, but want more attractive visualizations than what a spreadsheet can provide, it might make sense to use an **Infographic** creation tool such as Piktochart or Visme.

If you're looking for a free or relatively inexpensive dynamic dashboard that will update in real time, Google is currently beta testing a tool called Google Data Studio. It claims to offer "pre-built data connectors" that will make it easier to feed external data into the system. And, as you would expect, it is seamlessly integrated with every source of Google data. You can create up to five dashboards for free. To create more, or to allow access to multiple report users, you can pay monthly, but pricing is not publicly available.

There's also a whole class of business intelligence software tools designed to help you pull together data from different systems and transform it into easily readable reports and dashboards. This includes tools such as GoodData and iDashboards. Tableau is a favorite of many nonprofits, partly because it's free for nonprofits with a budget under $5 million, but also because it's powerful and easy to use. Other large software packages include the nonprofit-focused JCA Answers, although it can only be used with a few database systems. In general, these tools are expensive. You can expect to pay $1,000 or more.

Another option is to use a *Custom Reporting Tool*, such as Crystal Reports or JasperReports, to create a dashboard for your organization. These tools also have the advantage of flexible custom reports, which provide more information than would be otherwise available.

# Maps and Geographical Information Systems

Maps and Geographical Information Systems (GIS) allow nonprofits to display, analyze, and share data such as addresses, ZIP codes, or latitude/longitude coordinates. By displaying information visually, maps can reveal significant data relationships that would otherwise be hard to notice. These tools range widely in complexity and in the features they offer.

It's critical to spend time understanding what information will be required, where you can get it, and the time it will take to transform it into a format that can be used by your (or any) GIS system. Pay close attention to ensure that your data is accurate—a misread GPS device or a misplaced decimal point can literally make a mountain out of a molehill. In addition to geographic data, you might also look to publicly available demographic information, such as ethnicity or income level.

For many nonprofits, simply providing flat-map or globe views of the world, through what are often called "geobrowsers," will be enough to make sense of program data. These tools—such as Google Maps and Google Earth, Map Builder, and Bing Maps—are relatively simple to learn and use, allow you to plot data to create basic maps, and let you share your maps, but don't allow for in-depth data analysis. Google Drive's spreadsheet tool or BatchGeo can help you load a whole spreadsheet of addresses rather than plotting them one at a time.

Think about the type of data and how you want to visualize it before picking a tool. For example, if you need to create thematic or "heat maps" to display election results by district, you can use a free tool called OpenHeatMap.

More sophisticated GIS packages allow you to work with data that includes imagery (such as maps), points (like a building), lines (such as streets), and polygons (areas enclosed by a shape, such as a census tract). You can then describe these points, lines, and polygons with other data, such as income levels, vacancy rates, or ethnicity. All of this data can be displayed on top of one another in layers on a map.

Sophisticated GIS allows more advanced understanding of all this data when layered together through functionality that includes queries and filters that help analysts focus in on particular data and layers; functions for thinning and generalizing data; tools for reconciling physical features from two different data layers into the same view; and more. Advanced tools such as MapWindow, Manifold System, or the industry-leader ArcGIS require strong data analysis skills, and using these tools effectively can require quite a leap in expertise from the basic mapping systems.

# Measuring Social Media

Once you start using social media, how do you know if your efforts are working? There are many tools and options for measuring your success, from simple general approaches to tool-specific solutions.

Most social media channels offer some built-in way to analyze your activity. *Facebook* Insights is a reasonably powerful tool that shows data on your overall page as well as individual posts; *Twitter* and *LinkedIn* offer the basics—impressions, engagements (likes, clicks, and retweets), and engagement rate. Facebook owns Instagram, so its analytics are similar. Most blogs can be effectively measured with a combination of the blogging site's built-in metrics and *Website Analytics*.

*Website Analytics* tools such as Google Analytics can help you monitor traffic to your website from all your social media channels. And Google Trends can give you a big picture view of how many people are talking about your organization. You can also use Google Data Studio to create reports that aggregate social media data and report on it in one place.

You can use free tools such as Social Mention or Addict-o-Matic to search for your organization or topics you're deeply invested in. Google Alerts can send you emails every time a specific term is mentioned on the web.

For more streamlined or automated monitoring, aggregating tools help keep an eye on multiple aspects of your social media efforts. Tools such as Keyhole, Klout, and Mention aggregate mentions of keywords or your brand across multiple social media channels—although they all charge monthly fees and can cost as much as $1,000 per year. Many of the social media management tools also offer useful analytics, including Hootsuite, Perch, and SproutSocial.

For larger organizations, NetBase allows you to monitor social media activity in 42 languages and Synthesio tracks niche social media sites across the world. At the higher-end of the price scale are Radian6 (now part of the Salesforce Marketing Cloud) and Lithium, which measure influence, deliver comparisons to competitors and data on your market share, provide a customer service channel, and offer tangible methods of improving.

# Online Listening

Online listening tools help nonprofits "hear" what people are saying about their causes or organizations online. Whether they're saying good things about your programs or questioning your methods, knowing what they're saying and who's doing the talking can help you fine-tune your plans, consider future actions, and prepare appropriate responses.

A number of online listening tools require you to actively search. You type in a keyword, such as your organization or executive director's name, to find online discussions, conversations, or mentions. Different tools used to be distinguished more by what online areas they search—for example, the entire web, a specific social networking site, or blogs—but the trend is to consolidate and search all the major platforms.

Addict-o-Matic and Social Mention are free tools that offer broad searches. IceRocket is a free tool that focuses on blogs and *Twitter*.

Similarly, you can search for topics in *Facebook*, *Twitter*, *LinkedIn*, YouTube, and Flickr, among others, by using each site's respective search functions.

If you're looking specifically for web content, you can use an RSS tool to create a "listening dashboard," although most RSS readers focus on the feed from a specific source rather than a feed by topic. Feedly offers a free feed that includes keyword alerts, although it's primarily using Google Alerts for this. Many of the tools in *Measuring Social Media* offer listening features that include a wide range of platforms.

Higher-end tools such as Radian6 (now part of the Salesforce Marketing Cloud) and Jive can create robust listening dashboards with less work, but they're best suited for organizations with the resources to afford them and enough online mentions to make it worthwhile. Don't forget the more traditional ways to listen to people, such as online surveys, phone calls, or old fashioned feet-on-the-ground conversations with people in your community. You can learn a lot about the effectiveness of your social media efforts (and get helpful advice) simply by talking to people.

## Consider a Clipping Service

Clipping Services, or Media Monitoring Services, can monitor social media as well as online news outlets and print media. Most offer software so organizations can track these sources themselves, but others can do the tracking for you and deliver reports on the results. Sprout Social focuses primarily on social media, but at higher tiers, there is Trackur, Critical Mention, and CustomScoop, which can cost hundreds or even thousands of dollars per month. These tools allow you to keep track of social media followers, note when media outlets promote your resources, or look at the ratio of positive-to-negative press about your organization.

# Program Evaluation

Articles, conferences, and books have been devoted to the concept and strategies behind evaluating the success of programs, but there's no single software specific to the practice. Program Evaluation is not primarily a technology issue—software can help you track and analyze the data you'll need, but in order to draw any meaningful conclusions, you need to first carefully identify your evaluation strategy.

Consider how complex your data needs are. Can you easily track everything you need about each point of data in a spreadsheet row? If so, don't shy away from a simple Microsoft Excel or online Google Drive spreadsheet.

Many nonprofits track much of their data at a client level and then roll it up across multiple clients to help them better understand the broader impacts of their programs. If the data you need for your evaluation plan is tracked at a person level, a Case Management System, Constituent Relationship Management (CRM) system, or some other kind of constituent database is likely to be the best way to manage and report on it.

If you're tracking things that aren't specifically related to particular constituents—for example, water quality over time, or the outcomes of 30 different programs conducted by 30 different organizations—start by checking to see if you're using any other systems that track information at the same level. If you're already tracking information on the 30 programs (in a grant management system, for instance), you can likely also store and report on the outcomes of those programs in that same system.

But if other options fail, you may need to use a database platform to build your own system. Consider Microsoft Access or FileMaker Pro, which give you a base set of tools on which to build systems to enter and report on data. Or consider a Constituent Relationship Management platform such as Salesforce or Microsoft Dynamics CRM. These platforms will let you track and interrelate both constituent and non-constituent metrics in the same system.

There are also tools geared toward helping you understand what your data means. ***Charts, Diagrams, and Infographics*** and maps and ***GIS*** tools can help you to visualize your data, while ***Dashboard*** tools can provide summarized views. If you're looking to provide statistical evidence of efficacy, you'll need statistical analysis tools to analyze quantitative data about your programs.

# Survey Data

Online survey tools are a cost-effective way to develop questionnaires, distribute your questions, collect results, and even analyze the data—all through one central package. These tools let you easily define survey questions and possible responses using an online interface, send constituents links to take the survey, and download response data. A number of dedicated tools are available, but basic survey features can be found in all-in-one CRMs, **Broadcast Email** tools, or online form builders such as Google Forms, Wufoo, or Formstack.

SurveyMonkey, SurveyGizmo, and Polldaddy offer free options with limited functionality. You can upgrade on each of these platforms or choose other services such as FormStack or QuestionPro for between $200 and $1,000 per year. The paid versions tend to provide considerable support for different types of survey questions, and analysis tools continue to improve. Some of them even let you embed a form in your website rather than redirect people to an external site.

If you want to conduct larger-scale research projects, a more powerful survey package such as Qualtrics, CheckMarket, or Key Survey might be a better fit. These tools, which typically cost a few thousand dollars per year, support more sophisticated question formats, survey logic, and data analysis. LimeSurvey, a free and open source tool, provides similarly advanced functionality. The more complex feature-set makes this whole class of tool more difficult to use without training—especially for those without prior survey-design expertise.

There are a few useful techniques for interpreting survey data that you might want to try. Your survey platform will likely create a report that counts the number of each kind of response. Some can even categorize comments with simple tags such as "positive," "negative," and "neutral." Anything the system counts for you is relatively straightforward to put into a graph and draw conclusions based on common responses and the relative differences.

If your survey includes demographic information, then you can further slice up your data by groups of people to help you better understand the nuances between their responses. However, be careful that you don't slice your data too finely. If you find you're creating subgroups with only one or two people in them, you're probably not getting representative data and are likely to draw uninformed conclusions.

Responses that cannot easily be counted will require additional work. Some organizations take an intuitive approach and simply read the responses and look for patterns that jump out at them. Others attempt to code and quantify the data. For example, a lot of respondents might have made comments about the quality of services. You can break those down by cost, outcomes, friendliness—however you want to categorize them. Then, within each category, you can create codes. For example, if cost is a concern, then you can create codes for "expensive," "inexpensive," or "just right." Then you would go through and label each relevant response with a code. It helps to do this in a spreadsheet to make it easier to count the number of times each code was applied. Once each response is coded, then you can quantify the data to find out how many times a sentiment was expressed and analyze the data further by finding out the percentage of those kinds of responses or how often one response was tied to another.

Essentially, although survey data is inherently subjective and qualitative, you can still derive useful numbers that you can then analyze to get a better picture of how people think about your organization or its services.

## Just a Few Questions to Ask?

If these tools seem like overkill for your needs, consider Google Form functionality, which lets you easily create and post a form online for free. The Summary Report feature provides a helpful executive report of data and submissions, too. Polldaddy offers similar features to integrate a small poll into your website.

# Website Analytics

Web analytics software tracks your site's statistics—visitors to each page, what sites they came from, who they are, what browsers they're using, and more—to help you understand and improve your website and readership.

You may already have some of the tools you need. It's possible that the vendor you pay to host your website offers you access to some web statistics through the same control panel you use to administer email addresses, check available file space, and manage permissions. These tools (AWStats and The Webalizer are common packages) offer basic reports with little in-depth analysis, but are a free, convenient way to get your feet wet with analytics.

Google Analytics is widely considered the dominant player in this space. It's free and widely used, and dramatically more powerful than any other free option. Getting started requires access to your website's HTML code, and at least a few hours of work— and more to track documents such as PDFs or Flash content. You'll be able to see sophisticated metrics, analyze data across timeframes or pages, and set up the "traffic reports" you'd like to see. The powerful interface may overwhelm less tech-savvy users, but there are a lot of books, training courses, and free YouTube tutorials available, including Google's Analytics Academy.

Piwik is a free open source option available for download (meaning you store the data on your own machine). It is generally considered less user-friendly than Google Analytics, but some organizations prefer the open source ethos and the fact that they hold the data, not Google. Piwik also offers a Cloud-hosted version starting at $29 per month and determined based on the volume of "actions" on your site.

A number of robust analytics tools, such as Clicky, KISSmetrics, and Chartbeat, offer more attractive **Dashboards** or the ability to zero in on where visitors drop off, but are not substantially different from Google Analytics. Many are also good at tracking sophisticated multimedia content (which Google Analytics can't easily track). They also provide technical support, which Google Analytics does not. Other tools such as Inspectlet may provide heat maps (a visualization of where users are looking or moving their cursor) and screencapture. Prices for any the tools above vary, starting as low as $10 per month and ranging up to $1,000 per month or more.

*Almost every nonprofit needs to coordinate the work of multiple people, from staff to volunteers to board members. Collaboration software can help you share information, hold conversations, and manage projects—whether your teammates are in your office or around the world.*

*All of the software names highlighted within the text are covered in more detail in this guide. Most of them are included in this section, and follow immediately after the descriptions.*

COLLABORATION

# Collaborating with Others

Even if you're a single-person organization, chances are you don't work alone—these tools can help you collaborate with colleagues, board members, volunteers, and consultants.

## Strongly Consider...

If your staff or team members are spread out geographically, or if you regularly meet with constituents, volunteers, or others, *Online Conferencing* can provide an easy way to facilitate meetings without the overhead and logistics of travel or accommodations. It can also make it possible to host trainings, Q&A's, and other events.

## Keeping Ahead of the Curve...

If you work with multiple people on multiple tasks and deadlines, *Project Management* software can make it easier to keep responsibilities and accomplishments straight and help people understand their roles in the grander scheme.

For some tasks, *Collaborative Documents* might be a better fit for your needs—they allow multiple people to work on a single document at the same time. *Screenshots and Screencasting* can also help with online trainings by creating images of your computer screen—or video recordings of your actions on the screen—to pass along to others by email or over the web. Screenshots can show how software looks, or is supposed to look, while Screencasts can teach others how to perform certain actions.

## On the Cutting Edge...

*eLearning* takes online training to the next level. With the right tools, you can create instruction modules rich with multimedia content to instruct students in an out-of-the-classroom setting. Such lessons used to require specialized course developers to design and create, but now they're within reach of anyone with the right tool and a good plan. *Learning Management Systems* are another option for organizations that offer full-fledged courses or need to manage training.

With *Online Chat*, people can type questions on a website to communicate with an audience—say, your organization hosts a Q and A with an expert, and constituents can log on and see the questions and answers the moment they're typed, or ask their own.

If your organization wants to collaborate on content to a greater extent than collaborative docs support, a *Wiki* might be the answer. *Wikis* are user-editable websites that serve as a comprehensive, easy way for a large group of users—like staff or constituents—to create and share a growing pool of information.

Finally, *Board Support Software* is growing in popularity in the nonprofit sector. This cohesive collection of tools and materials enables members of an organization's Board of Directors to access, print, and comment on board documents, take notes, and communicate among each other and with staff, keeping board members informed and engaged.

# Board Support Software

When board members and staff are spread across multiple offices or geographic locations, collaboration is not as easy as sitting around a conference room table. Whether voting on proposed budgets, preparing financial documents or recommendations for upcoming meetings, or evaluating pending grant proposals, sharing documents by email or participating in conference calls is sometimes not enough.

Software can bridge the distance and unite dispersed collaborators by making it easy to present, review, and comment on information. Some tools are designed for more general collaboration and can be tailored to meet the more specific needs of a board. Others are purpose-built for the board environment. Choosing the right one is a matter of requirements and budget.

Needs vary, but many boards share a number of tasks and requirements. These are the minimum features you will need:

- *Creating board books and other meeting documents.* Assembling a board book and related meeting materials is among the most important roles of a board collaboration tool. Because you may need to create, upload, and organize hundreds of documents for every board meeting, it's essential that the system make this process as smooth and practical as possible.
- *Sharing and accessing meeting documents.* A lot of work goes into creating board books and other materials—making it easy and intuitive for board members to read those documents is essential. In addition, they need to be able to annotate or mark-up meeting documents, highlight or strike-through text, and write notes or comments. Documents should also be easy to search through and navigate.
- *Sharing and viewing calendars.* Board members are busy people. They want to be able to easily see their meeting schedule and RSVP for board and committee meetings. Shared calendar tools such as Google Calendar and Microsoft Outlook do a reasonable job of this, but most board portals include built-in calendar tools.
- *Collaborating as a board or committee:* A board collaboration solution should enable members to work together on documents or committee work online. A dedicated board

portal will also support collaboration during meetings using online functionality for official voting and more informal straw polls for decisions that don't need to be recorded as part of the minutes. More full-featured systems may even include online conferencing functionality for fully virtual meetings, although many boards make do with Skype, Google Hangouts, or another videoconferencing tool.

For smaller organizations, dedicated board collaboration tools are not likely to be practical or affordable. You can create your own collaboration solution using a combination of file sharing systems, Google Docs, project management tools, shared calendars, and online chat or conferencing tools. This sort of makeshift solution can be a great way to get started quickly, or to meet less-complex needs, but doesn't provide as many features.

## Adapting File Sharing Systems and Project Management Tools

An online Cloud storage service allows you to share files with individuals outside your office. Many organizations with multiple locations or distributed staff already use such tools, including Dropbox and Box.com, to provide file storage and sharing. Google provides a variety of tools (Drive, Calendar, Groups, and Forms) that, used together, can meet many of the needs of board members.

General project management tools incorporate basic project planning, document sharing, task management, shared calendars, and online discussion boards into a single system. These web-based tools are particularly useful for geographically diverse teams or teams that include members from outside the organization. Huddle bridges the gap between file sharing services and more full-featured project management solutions. Basecamp is one of the most widely-used online project management tools and provides an intuitive interface and thoughtful email integration that allows individuals with varying levels of technological comfort to use the system. Sharepoint inspires strong feelings among users—both positive and negative—but continues to be a popular tool for document storage and sharing, and can be put to use for board collaboration purposes as well. Trello provides incredibly simple user interfaces with drag and drop functionality.

## Using Dedicated Board Portals

Board portals are designed specifically to help staff manage and create documents, and to help board members share, read, and annotate board books and other meeting documents through a user-friendly interface. As most of these tools originated in the for-profit sector and are designed for banks and corporate boards subject to stringent regulations, they are first and foremost secure.

In the past few years, prices have come down, drawing interest from the nonprofit sector. However, the tools remain expensive for many nonprofit organizations. Almost all provide streamlined, easy-to-use interfaces for tablets (primarily iPads) and robust permission and security settings. Many provide chat and notification functionality to allow board members to discuss duties and collaborate on meeting documents within the system. Few provide built-in online conferencing or other presentation tools, which means board members who cannot attend meetings in person will have to seek third-party conferencing tools.

BoardPaq is an easy-to-use tool providing strong support for basic needs such as document management and creating and sharing meeting documents. Passageways OnBoard provides a streamlined, easy-to-use platform with solid functionality for creating and reading board books and other meeting documents on both the iPad app and web interface. BoardEffect provides solid support for creating and managing board books and other meeting documents with an easy-to-use and intuitive iPad interface that allows board members to read and annotate documents. Diligent Boardbooks provides solid functionality for organizations that need granular permission settings or editorial controls over their meeting documents. Boardvantage provides solid support for creating and managing board books and other meeting documents. Lower-cost options in this space include MyCommittee or BoardMax.

Very few board portals can be considered all-in-one solutions. That's not to say that an all-in-one tool for board collaboration is necessary, or even desirable. Every organization's needs are different. Picking and choosing the individual software that best meets your needs presents the opportunity to customize a solution to your particular demands and budget.

# Collaborative Documents

Real-time collaboration on documents is difficult without a robust *File Sharing* solution, since files must be hosted online before multiple users can edit them. This requires a good connection to the internet and software that enables live editing.

Depending on the software, multiple users can make edits in real time, and all users see the changes as they are being made. Collaborative editing software also tracks changes and makes it possible to revert to previous versions. It also lets users leave comments and chat about a document, for example to suggest a change or ask a question.

G Suite (formerly Google Apps) first made these functions widely available through Google Docs and Sheets. Microsoft has begun to catch up with Office 2016. Other productivity suites, such as Zoho, provide similar tools.

Increasingly, *Online Chat* and *Project Management* tools build document collaboration into their communication and editing functions. For example, Adobe offers Creative Cloud to host and share publications, artwork, video and other large files, and Document Cloud, a free service for combining, commenting on, and signing PDF files. It is useful for overseeing business processes, such as signing off on an invoice or approving a reimbursement.

For larger projects, such as multi-chapter documents, you might look at a *Wiki* tool, which can provide nested pages instead of simple files-and-folders methods to organize content.

# eLearning

Online learning software pulls together such elements as slide decks and PDF documents, lecture videos, interactive on-screen content, and quizzes. The result can be greater than the sum of the parts if it's done correctly. Successful course authors must mix the instructional experience of a teacher, entertainment instincts of a film director, and insights into user experience of a web designer.

Software can't make you a good course designer, but it can provide the tools to distribute the lessons.

Adobe Captivate, TechSmith Relay, and Articulate are a few examples of eLearning tools that let users embed content created in other programs such as PowerPoint and videos, and add text, shift elements, and more. Adobe Captivate integrates access to its stock images and video collection, although licensing images may entail additional cost.

Versal seeks to provide a simpler interface to create courses, letting users select and customize features from a long list of "gadgets." Another program that takes a slightly different tack is Lessonly, designed to crowdsource the instructional side of course development—for example, employees sharing knowledge about their company to help create a course "onboarding" an organization's new hire.

Many programs include proprietary ways to distribute the finished course. However, they usually can export content to a standard file format, Shareable Content Object Reference Model (SCORM), that can be used in any *Learning Management System*.

# File Sharing

Email attachments are still many users' default method for sharing files. But file sharing is an essential replacement to move big files, such as video, audio, or complex presentations. It's also a good first line of defense for *File Backup and Recovery* and, done right, can be part of a *Document Management System.*

You can navigate to shared files on your computer just like other folders on your internal hard drive or office server. But because files you share live in a folder stored remotely on the internet, you can create a link to a large file that anyone can use to download the file, and even choose to share them publicly on the internet.

One challenge to file sharing is the need for users to understand about access and permissions. After creating a file but before sharing it, authors must grant access so others can see it, and set the appropriate permission level: view, comment, or edit. Further, some file-sharing tools allow a file owner to decide whether others will be able to invite others to view or use it.

If your organization is just getting started with Cloud-based software, then productivity tools like Microsoft Office 365 and G Suite (formerly Google Apps) can be a good introduction to file sharing. For example, 1 TB of Microsoft OneDrive storage is bundled with Office 365, and G Suite gives each user 30 MB of data storage.

More robust file sharing is available from Box, DropBox, and SugarSync, among other services. All three provide apps to sync files across devices, so you can access the latest version on a phone, tablet, or computer that is connected to the account, regardless of where it was last edited.

Box, DropBox and SugarSync offer free "individual" file sharing versions that offer a single user limited amounts of data. However, they specialize in providing teams with increased storage space and more tools. For example, Box offers a Starter account designed for teams of three to 10 people via Techsoup. It costs $84 a year and includes 100 GB of data and a limit of 2 GB on any single file.

For $750 per year (less with a nonprofit discount), DropBox Business gives five users unlimited space. In addition to access and permissions, DropBox Business includes services like remote wipe in case of theft or loss of a device. SugarSync provides an equivalent service for slightly higher cost (five users, 1 TB, $972). Nonprofit pricing is available upon request.

These and other file-sharing tools add features to help them integrate smoothly with your operating system and software. For example, Egnyte lets users edit files directly from in-the-Cloud remote drives. Business editions of Box add features that support document management, such as the ability to tag documents so they will be easier to find later.

Team file-sharing tools also let administrators transfer ownership of a set of files from one user to another, and remotely wipe files from one or more devices. This can be helpful in case of loss or theft of a device, or when a staff member leaves the organization.

A final note about file sharing. There may be times when you just want to use a file-sharing service to send one big file to someone else. The services already mentioned here may be the best choice, but as a backup, consider tools like pCloud Transfer or Transfer Big Files. These allow users to send files that are 2 GB to 5 GB in size without a Cloud-based file sharing solution. However, the speed of sending large files this way will depend on your internet connection.

# Intranets and Portals

*Websites* are ideal tools for sharing your story with the world, but leave something to be desired when it comes to internal communications—namely, privacy. An intranet or portal site is a password-protected site that can be hosted on the web or on your office server (thus the name "intra" net).

Like a website, you must think through the business purpose of an intranet when designing it, and build in resources for training and ongoing maintenance. Common functions include employee directories with contact information, in/out of office notifications, and announcements. A staff intranet might provide HR documents such as reimbursement forms or manuals.

Another variant is a company "extranet." For organizations with affiliates or volunteers, an intranet or extranet portal can provide a private site to share resources, access to members-only information, or tools relevant to their work with your organization.

If you already have a website built in WordPress or Squarespace, you can find a compatible way to add a members-only intranet space through add-ons such as SimpleIntranet for WordPress and MemberSpace for SquareSpace.

If you use the Salesforce CRM, you can find the Community Cloud app on the Salesforce app exchange. And Microsoft recently unveiled new ways to use Sharepoint to create an intranet that can be accessed via smartphones and tablets as well as on a computer. But all of these options require some technical expertise to create a polished, well-functioning site tailored to your needs.

Another approach is to look at purpose-built Intranet tools. Examples include Papyrs ($2,400 per year less a 10 percent nonprofit discount), Interact Internet ($96 per user, per year, up to 100 users), and Liferay. The latter presents users with the ability to assemble functions from an app store, so it can be personalized.

# Learning Management Systems

*eLearning* tools let you create courses, but you'll need a *Learning Management System* (LMS) to register students, track their participation, and let you know when they complete a course. Since many classes incorporate discussion with an instructor, some form of video conferencing may be needed, too. These and other add-ons are some of the variables that differentiate the many LMS products available.

Any LMS should be able to deliver the courses you acquire or create, because eLearning tools use standards called SCORM (Shareable Content Object Reference Model) and xAPI (Extensive Application Programming Interface), both developed by the U.S. Department of Defense to insure interoperability of training courses.

Some of the most widely used LMS tools are open source and require either specialized knowledge or engaging an expert partner to customize, including Moodle, Totara, and Sakai. Moodle offers a free version that lets up to 50 people take a course and offers 200 MB of free storage.

You will find big brands as well as niche players providing LMS tools. Tools like Blackboard and Scholar LMS focus on multiple features related to delivering course content. However, some tools offer a companion proprietary eLearning tool. For example, Adobe offers Captivate Prime LMS alongside its Captivate software, and iSpring offers both iSpring Suite for authoring and iSpring Learn LMS. A few companies, such as Litmos, go a step further and provide courses you can use for your organization in addition to authoring and LMS tools.

# Online Chat

Online Chat combines elements of a phone call and email exchange, in that it allows for the immediacy of the former and the written precision of the latter. Originally, online office chat functioned a lot like text messaging, but using desktop computers rather than mobile phones. Then group messages emerged, and later still, developers borrowed popular features from Twitter and Voice Over Internet Protocol, or VOIP, letting users start live phone calls or video conferences from the Online Chat tool.

Slack, HipChat, and Campfire (by the same people who make Basecamp *Project Management* software) add *File Sharing*, *Document Management* systems, and *Project Management* functions to their online chat tools, and HipChat puts screen sharing at the core of its tool, among many other capabilities. Slack features an ecosystem of hundreds of add-in apps by third-party developers for everything from project management to time tracking to unique sets of emoji. Facebook offers one similar to its Messenger tool called Workplace by Facebook. Microsoft recently rolled out Teams for Office 365. All of which is to say, there are no standardized features to online chat tools, and each one offers something a little different.

These tools tend to be low cost or free to nonprofits. Both Slack and HipChat have free-to-anyone basic versions and are free to eligible nonprofits (Slack standard edition is free for up to 250 people to nonprofits and sold at an 85 percent discount after that), as is Workplace for Facebook. If you already use Office 365 at work, Microsoft Teams just needs to be switched on in the admin portal. You can also use Google Chat if your office relies on Gmail.

One other area where nonprofits are beginning to add chat is on their websites, a technology most commonly seen for customer service applications. Tools like Livechat or Clickdesk work by inserting a bit of HTML code on your website to add a floating chat button that lets visitors to your website reach out to you with questions. If they contact you outside work hours, these tools allow the site visitor to "leave a message" for you to reply to when you return.

# Online Conferencing

Technology has made it possible for the workforce to spread out geographically, but email and phone calls only go so far facilitating communication. Sometimes you need visuals to illustrate a conversation. Whether you want your audience to see the same slides or document over the web, display your computer screen for a demo, or conduct more formal online seminars (often called *Webinars*), Online Conferencing tools can help.

If all you need is the ability to conference by video, Skype, ooVoo, join.me, Freeconferencecall.com, and Google Hangouts offer useful and free audio and video conferencing for those who have a microphone and webcam. All of these services also allow you to share your screen, but some display ads during your conference. WebHuddle, an open source option, is also currently free. Audio and screen-sharing quality with these tools can vary and is typically less reliable than with the more expensive tools below.

If you need to share your desktop or an application with your conference participants, a number of low-cost tools are available. Yugma allows you to share your screen, share screen control, collaborate on a whiteboard, and take comments by chat for between $25 and $1,600 per year based on the size of your group. Skype for Business is $5 per user, per month and allows you to collaborate on Microsoft documents such as PowerPoint presentations and Excel spreadsheets (if users have Office 365). Join.me offers a business version that gives more control over users and features for $25 per month. UberConference is another affordable option at $10 per line, per month. TeamViewer can be used for conference calls, but specializes in remote IT support because it enables a remote user to take control of a device. It is also pricey at $849 for one license and unlimited participants.

Tools such as GoToMeeting, GoToWebinar, WebEx, Adobe Connect, and ReadyTalk often cost more, but are more widely relied on and dependable. They often provide such features as integrated audio and visual recordings, integrated toll-free conference-calling lines, scheduling features, and other advanced interactive tools. Cost is often complicated to determine, as it's based on the features you want and the number of people in your meetings. It generally ranges from $200 to $5,000 per year.

# Check on Audio Conferencing!

Note that the posted prices for these tools often do not include the cost of audio conferencing via a phone line. Some tools offer Voice Over IP (VOIP), which provides free audio over the internet, but that requires all participants to have computer speakers as well as some technical know-how. Otherwise you'll need a telephone conferencing line. Online conferencing vendors will generally provide one—often at extra cost—or you can use a service such as FreeConferenceCall.com. Note that FreeConferenceCall.com is free for you to use, but it does not provide a toll-free number, so participants in your conferences must pay any of their own long-distance services to call in.

# Project Management

Simple projects can easily be managed in a spreadsheet program like Excel. But if you have ever created an elaborate spreadsheet for a more complex project, with columns for tasks, dates, responsible team members, deliverables, and more, you know how unwieldy it can become. You might be ready for Project Management software. Born in industries like construction, defense, and technology, software engineers designed Project Management tools to mesh together processes with lots of moving parts.

These systems offer users a dash of **Online Chat** by providing updates and "likes"; a smidgen of **File Sharing** and **Document Management Systems** to save, tag, and categorize key documents; and a bit of **Intranets and Portals**, such as team timelines and shared calendar for all team members. Time tracking (useful for reporting against specific grants, or clients), budgeting, and other features may be available, as well. The sum of these parts is a single tool to help organizations define goals and deliverables, plan who will do what and by when, and steer project-related communication away from each team member's email inbox. Experts recommend they be used to organize work toward specific goals, as opposed to day-to-day tasks.

Virtually all **Project Management** tools allow a free trial and are relatively low cost. Asana provides a free "lite" version for up to 15 users. Basecamp is free to try for an unlimited time, but provides increased storage for paid subscribers. Confluence is free to nonprofits (although only for an on-premises version). Trello follows a similar model, but the paid version also includes more features. Zoho Projects lets unlimited users work on a single project for free. All offer mobile apps to access projects outside the office.

While Asana, Basecamp, Trello, and Zoho Projects all operate in the Cloud and come with varying levels of Cloud-based file storage, project planning software is also available for local installation on your own server. This avoids per-user charges, gets around the need for Cloud-based storage, and may work well for organizations primarily housed in a single office or reluctant to go all-online. Active Collab, Confluence, and Microsoft Project are available to install either on premises or in the Cloud.

The best project management tool may be an add-on to a tool you already use. Microsoft Sharepoint comes with project tasks lists built in, and offers many third-party tools in the Microsoft Store. GQueues is a third-party tool that integrates with G Suite. If your organization uses Salesforce, several project management extensions are available via the Salesforce app exchange, such as Milestones PM, or Mavenlink Project Management.

If this area is of interest, you may want to look at one of the many courses available in project management, or even consider becoming certified through the Project Management Institute.

## What About Time Tracking Tools?

You may also want to explore lightweight, Cloud-based time tracking systems like Toggl or RescueTime for hour-by-hour breakdowns of how your team is spending its time on projects. This data is useful for planning out how much time future projects will take, as well as for identifying areas where staff time can be saved. These systems vary in sophistication of the data you can extract, but many have low-cost subscription plans that offer more features.

# Screenshots and Screencasting

At their simplest, screenshots are still images of your computer display that you can share with other people, and screencasts are video images of your display (for example, to create a how-to video for using a piece of software). They let you walk viewers through an online process, by adding text to an image or narration to a screencast, and they provide editing tools such as arrows, ways to fade or edit out parts of an image, and text boxes to highlight tips or instructions for a particular screen or workflow.

Windows and Apple computers allow you to capture the screen through keyboard shortcuts (Alt + Print Screen on Windows, Apple key + Shift + 3 or + 4 on a Mac). But dedicated tools for screen sharing and screencasting amp up that process, especially the ability to edit the captured screenshots and share a final product with others.

Techsmith's free platform, Jing, is good for capturing, editing, and sharing images up to a five-minute-long video. For a fee, the company offers Snagit, which adds functionality like additional tools to add text or pointers to an image and the ability to turn a screencast into a .gif. Techsmith also makes a screencaster, Camtasia, which captures video of a workflow process, for example. Snagit and Camtasia can be purchased at nonprofit rates of $50 and $180 respectively, or $204 together (nonprofit pricing).

Another company creating image capture and screencast tools is Blue Mango Learning. It, too, offers a less-costly quick solution, Clarify ($30) and more robust Screensteps ($99 per month).

Another common purpose of screenshotting is to document web research; for this purpose, consider web-clipping tools. Evernote, which has a free basic edition, lets you highlight text on a web page and "clip" the content into a note for later use, using tools like tagging to categorize items. You'll find many free and low-cost choices for this kind of screenshotting, including extensions for your browser, such as Awesome Screenshot for Firefox or Chrome.

# Wikis

A Wiki is a web page whose content can be modified by multiple people. (Perhaps the best known one is Wikipedia.) Administrators manage who can see and edit sections or pages. Wikis are great for creating content collaboratively. They can be accessed online from anywhere, they let multiple people edit simultaneously, and prior versions are automatically saved and easily restored.

The experience of working in a Wiki is more like working on the backend of a website, though, which can make updating one feel technical. For instance, staff members may need to learn and use formatting tags—for example, [h1] to denote that a header should be large and bold. These tags aren't particularly complicated, but can be intimidating to less-technical users.

PBWorks offers a free basic tool with features such as built-in *Project Management* tools. For more users, greater storage space, and more complex installations, the cost is $1,995 per year. Wikispaces offers relatively low-cost individual and small-team Wikis, with a more robust organizational version for a comparable price.

In addition to these tools, MediaWiki (the platform on which Wikipedia is built), Dokuwiki, and Tikiwiki are free and open source software packages. As with other open source tools, there's no cost to acquire them but they require specialized knowledge or an IT consultant to install and run.

## The Many Uses of Wikis

Wikis can be useful for creating or sharing information among a group—for example, as a collaborative website like Wikipedia, for project management or document development, to share best practices, or as an _Intranet_ or _Portal_. If your constituents are spread out across the country or the globe, a Wiki lets people keep pertinent data up-to-date in a central, easily accessible location.

THE FIELD GUIDE TO SOFTWARE

You have constituents: donors, event attendees, volunteers, board members, and more. You need a system to track them and to store all the information you'll need to build your relationship.

What type of system will best help you depends on your specific needs. This section will help you think about what exists, and what might work best for you.

All of the software names highlighted within the text are covered in more detail in this guide. Most of them are included in this section, and follow immediately after the descriptions.

CONSTITUENT MANAGEMENT

# Constituent Relationship Management

**At the heart of any Constituent Relationship Management system, or CRM, is a "relational" database—a database that connects tables with different types of information to an individual constituent.**

One table may hold data about donation dates and amounts while another has contact information and a third records event invitations and attendance information. Because all those tables are linked, you can search from either end—finding all donors who contributed to a particular campaign, for example, or finding out which campaigns a particular donor has contributed to.

Organizing different types of data in linked tables lets you aggregate information about your base, which can reveal useful patterns in their interactions with you. A simple example might be to review donation data to spot who responded to a successful appeal, then tailor a custom follow-up message to those donors— with messages going by mail to those who sent in a check, and by email to those who gave online.

That combination of donor data, contact information, and broadcast email is exactly the kind of 360-degree look at your constituents a CRM is designed to enable. But a CRM also helps you segment your list, tailoring messages to sub-groups of constituents to increase their engagement with your work and organization. Even simple segments, such as people who have opened every email you sent in the past few months compared to people who haven't opened any email in the past few months, can help you better understand and target your audience.

To these relational tools, add a graphic user interface that makes it easy for non-programmers to enter and review data in your database and you have the core of most CRMs on the market. On top of those features, some add other elements, from such basic

*Project Management* tools as setting up tasks with due dates related to a fundraising campaign to bits and pieces of *Broadcast Email* tools, *Credit Card Processing*, *Dashboards*, *Document Management Systems*, *Survey Data*, and even some web *Content Management Systems*. Which one you choose depends on your particular needs.

There are a number of different types of CRM systems available, including the following:

## All-in-One CRM

These systems are like the Swiss Army knives of CRMs. They're typically flexible to let you tailor them to the processes you need to support and, because they have mostly been around for some time, offer solutions to most common (and many not-so-common) nonprofit data-tracking challenges. If constituents have complicated relationships with your organization, or tend to cross organizational boundaries—for example, volunteers often become donors—a CRM can be a useful way to get a full picture of each constituent. Some people use "CRM" as a catchall term for any constituent database, but we use it to describe this specific kind of flexible system that has substantial integration with your organization's online presence.

## Advocacy-Oriented CRMs

Data has come to play as big a part in elections as in other worlds and a class of CRMs has evolved to address specific aspects of both voting and elections and civic engagement and advocacy.

## Case Management

*Case Management Systems* will track the information you need to work with a client, such as address, job history, medical history, and child care situation. They'll also track communications between your staff and the client, the individualized plan for your client, and the progress toward the plan, and let you report on all the information you've collected and maybe go so far as to support overall program evaluation.

## Donor Management

The annual Giving USA report has repeatedly shown that not only are individuals the largest source of philanthropy in the country, but also that individual contributions continue to grow year over year (with rare exceptions). These systems focus on delivering the tools you need to track donors, prospects, pledges, premiums and giving levels, matching gifts, and sometimes grants and corporate sponsorships, too.

## Legal Case Management

Tracking clients and evaluating programs are the tasks of *Legal Case Management Software*, which can also record communications between your organization and the people you serve, and more.

## Volunteer Management

Software has developed to help with every aspect of working with volunteers, from finding help—including qualified experts to provide legal assistance or assist with key tasks like back office and communications projects—to letting them pick the times when they want to help, and managing engagement for more complex or ongoing relationships.

## Specialized and Custom Constituent Tracking Systems

Some organizations need specialized tracking systems, such as churches and synagogues or those that want to track attendance at a lot of events. In practice, you'll see that despite the distinct purposes and designs, the different flavors of CRM we profile have much in common. Most offer a lot of support for donors and donations, track members and volunteers, and provide flexibility to track other constituents, plus some support for *Broadcast Email* and *Online Donations*.

Whether it's a Donor, Member, Volunteer, Advocacy, or All-in-One CRM system, it probably doesn't matter—as long as you find a tool that is sufficient to meet your needs.

# CRM Strategy Questions

In practice, CRM often describes a software strategy and set of processes as much as it does a specific class of software tool. Start by describing your constituents, and the processes you use to reach and serve them.

How many constituents do you have, and how big do you expect your database to grow in the next few years? It's common for groups to fall into categories such as under a thousand, a few thousand records, up to databases with 50,000 to 100,000 and on up from there.

What data do you already have about them, and what do you need to know? First name, Last name, Email address, and Zip code are a given—or should be. You need at least those fields to provide basic personalization and connect appropriately—for example, inviting only those who live nearby to a small planning meeting. Add street address for donors you plan to send end-of-year appeal letters by U.S. Mail. Add event participation or other key fields for program participants. Mobile and other phone numbers allow for direct contact, and ever-more-important text messaging.

Consider what features you will want your CRM to include. For example, many CRMs include a _**Broadcast Email**_ tool or integrate with a solution like MailChimp or Constant Contact. Similarly, many will offer their own _**Credit Card Processing**_ and _**Crowdfunding and Peer-to-Peer**_ solutions, while others will integrate with external tools.

Don't forget about data migration. You will likely face a cost to transfer data from whatever system you use now (even if it is just a collection of spreadsheets) into a new CRM. Someday, even if you are happy with the first CRM you select, your needs will change and it will be time to export all your data out to a new platform. Plus, imports and exports are essential to keeping data fresh and sharing it in reports

(continued)

and other forms. Be sure to know how you will handle these imports and exports.

Another question is how many users will access the CRM platform, and how many need to be in it at one time. Some systems provide a limit on how many users can be in the system, or how many can use it at one time.

The more people in your organization who embrace the CRM and use it, whether as their online address book, to track work and assign meetings and tasks to colleagues, or send emails, the better the quality of the data is likely to be. How much training and support will you need to get to a critical mass of users?

You should also consider whether your CRM will be hosted locally at your office, on a remote server you connect to via the internet, or in the Cloud. The different options can make a big difference in the ultimate security of your data; just like webhosting or Cloud-data services, consider what protections and guarantees of backup and uptime you need.

Finally, know what other programs integrate with any CRM you choose. If you plan to take money via your CRM at all (and you probably should), you'll need to understand how the database and any associated payment processors connect to your accounting software.

Integration is also important if you want to keep using your existing broadcast email provider and credit card processor alongside your new CRM. Many organizations use this kind of strategy to control costs and get some of the same 360-degree understanding of their constituents that a single CRM provides. Well-thought-out CRMs typically will offer an Application Programming Interface (API), a standard software feature to facilitate interaction with other software and systems.

# Advocacy-Oriented CRMs

Civic engagement and political campaigns have led to new platforms to complement other *eAdvocacy* tools. In addition to fundraising, broadcast email, and petition tools, these CRMs focus on helping you and your constituents organize and take action.

For example, they focus on integrating social media tools your constituents can use to spread the word about their involvement with your group and draw in others. They incorporate or integrate with tools to send a letter directly to an elected official (using built-in datasets with electeds' up-to-date email addresses) and other features, and can be customized to match your website branding.

They also offer access to state voter files and protocols to insure donations comply with Federal Elections Commission rules if you need those features, for example for a 501(c)4 organization. Because these platforms are of relatively recent origin, you can expect user-friendly visuals and a clear graphical interface.

Action Network, itself a nonprofit, offers "lite" free accounts; the paid version can scale up to much larger deployments. BSD Tools, named for the company that created it, Blue State Digital, starts at $6,600 for up to 50,000 records. The price includes services that might be add-ons with other CRM platforms, such as text messaging.

If you've used the voter-files database Voter Activation Network (VAN) then you may already be comfortable with the advocacy CRM EveryAction, by the same firm that developed VAN. Pricing starts at about $600 a year. NationBuilder starts at $240 a year for 1,000 email-able individuals or $1,600 a year with features such as fundraising, events and ticketing, and text messaging.

# All-in-One CRMs

If the idea of investing in a single system everyone on your team uses for just about everything across the organization appeals to you, then one of these systems is likely a good fit.

Organizations with smaller datasets may find the cost of installing and maintaining an all-in-one solution comparable to a specialized tool. But medium to larger organizations may find more benefit from comprehensive all-in-ones than smaller groups. They have staff to take advantage of the features available in these platforms and the ability to make the most of them through custom installations.

By design, all-in-one CRMs handle lots of data and relate very different data-sets to each other. An individual's record might offer data on their petition signatures, e-commerce, survey responses, and relationships to others in the database in addition to standard data such as contact information—not to mention any custom data your organization has chosen to collect.

In addition to offering lots of processing power, all-in-one CRMs are designed to allow for customization to reflect the nature of your work. Chapter-based national networks, think tanks with lots of publications to share, large social-service organizations with a need for integrated case management—these are just a few use cases for an all-in-one CRM, and each would benefit from, if not require, a custom build-out.

All that power likely comes with a layer of complexity. Perhaps because they have so much going on, graphical interfaces of all-in-one CRMs may be less user-friendly than other systems. Customization can help with this by adding a layer of polish, and providing users with training so they know how to navigate the system.

Each of these CRMs has spawned its own community of experts

and consultants who can provide the necessary services to implement and maintain your CRM. If you need a function that doesn't yet exist for the all-in-one CRM you select, a developer may even come forward to design a specialized tool for you.

As free and open source software, civiCRM is available to anyone who wishes to download it. But installing, hosting, and maintaining it takes substantial knowhow and infrastructure, both of which have a cost. As an example, a civiCRM partner might charge $400 to install the software and about $1,500 per year to host, maintain, and provide support for a database of up to 15,000 records.

Several well-known all-in-one CRMs are nonprofit offshoots of products developed in the for-profit world. At the core of the Salesforce.org program for nonprofits is a donation of 10 user seats, or licenses. Additional seats cost about $430 per user, per year. As with civiCRM, "free" may come with costs for a consultant to customize the CRM and train you and your organization in its use. NetSuite, whose Enterprise Resource Platform is typically used by larger organizations and corporations, also provides nonprofits with 10 free seats.

Neon CRM and Salsa are examples of all-in-one CRMs developed for the nonprofit sector. Blackbaud, which makes several different CRM platforms for nonprofits, is discussed in detail in the section on **Donor Management Systems**. Additional products include Microsoft Dynamics, SugarCRM, and, Zoho, although the latter two are designed more for businesses out of the box.

Perhaps worth noting is that these systems are called all-in-one because in addition to core functions they offer add-ons to do the same things as the specialized databases described elsewhere in this section. For example, when it comes to advocacy, Salsa has a built-in petition tool that is widely admired; for **Volunteer Management Software**, civiCRM offers CiviVolunteer, and Salesforce's app exchange includes the Volunteers app for Salesforce.

# Case Management Systems

Case management systems—sometimes called client management systems—will track the information you need to work with a client, such as their age, address, job history, medical history, and child care situation. They'll also track communications between your staff and the client, the individualized plan for your client, and the progress toward the plan, and let you report on all the information you've collected, maybe even including overall program evaluation.

Advanced case management systems can do even more, such as helping with workflow and scheduling. For instance, based on the information you enter about a client, they can recommend that your client meet with a dietician, help to schedule that meeting, and send a reminder. They can also help to automate your billing processes.

A number of systems are intended to work across a range of human service scenarios. More generic systems such as Exponent Case Management for Salesforce or Social Solutions' Apricot can provide substantial tracking ability and flexibility to support existing intake and service workflows for $3,000 to $25,000 per year. Somewhat more expensive software, like Social Solutions' Efforts to Outcomes (ETO), Bowman Systems' ServicePoint, ClientTrack from ClientTrack, Inc., Service Xpert Suite by Unicentric, or Tapestry by Visionlink, can provide more power for more expense—about $20,000 to $50,000 per year. Many are designed with particular best practices in mind and require users to follow their particular processes and workflows.

Those looking to integrate client information with data about their other constituents (like donors), or to support complex and unusual processes, should also consider **Constituent Relationship Management** (CRM) systems. These systems tend to be very flexible, but provide less specialized client-tracking functionality out-of-the-box. For instance, a number of organizations are adapting Salesforce for use in case management, or using CiviCRM's CiviCase component. If your programs concentrate on a particular, widely recognized area, such as homeless management, child care, health services, summer camps, or legal aid services, consider tools geared specifically to your type of work.

# Donor Management Systems

At their core, donor management systems connect detailed information on each gift to a robust dataset about its giver. They also offer several distinctive components. Foremost is the ability to handle both gifts and pledges, including recurring gifts. Another tool you can expect is the ability to tailor your donor management system to reflect the appeals, campaigns, and funds that you use in your day-to-day fundraising work.

Forms that integrate with your website are another useful feature, as research shows more donations come on branded donation pages that resemble your organization's website compared to third-party donation pages. With so many prospects browsing the web on smart phones and tablets, those forms should be mobile-friendly. Systems may integrate with other **Credit Card Processing** tools to take payments via card or ACH (bank account) transactions, or offer their own Payment Processing Systems.

Support for **Crowdfunding and Peer to Peer Fundraising** are common tools in donor management systems. Other tools you should expect include **Broadcast Email** to allow for online communications, appeals, and follow-up, including automated gift receipts and thank-you notes. This may be part of the software or rely on integration with a third-party tool, such as Constant Contact or MailChimp. The ability to write thank-you letters and store these and other correspondence with donor records is another useful feature.

If individual contributions, including major gifts, make up a good part of your organization's revenue stream, consider software with functions to track your organization's every engagement with a donor or prospect, connect to external helpers like wealth screening tools.

**Dashboards** are another tool to help you make sense of all the information your system holds. These graphical interfaces use charts and numbers to snapshot your results. Common metrics to look at include percent of donors who have given previously (retention rate), total gifts, and total raised—all good numbers to know for planning and other purposes.

If all this sounds complicated, then one more feature that some donor management systems offer may be the most helpful of all—built-in coaching with an experienced fundraiser who can help with implementing the system for its goal: raising funds for your organization.

In selecting a donor management system, you should first take into account the size of your database and complexity of your needs, then weigh your desire for simplicity vs. cost. If you're just getting started, check out the options available via TechSoup; several firms offer one free or discounted year on their donor management systems.

With some exceptions, you can expect to pay $2,000 a year or more for these services. Among the best values for organizations starting out or that have a small list—fewer than 1,000 records—is Little Green Light, which offers a basic package at $425 plus discounts for new users via TechSoup; it's also economical for larger lists.

Many donor management systems offer economical basic packages and let nonprofits select add-on features for an additional cost. These include DonorPerfect, which offers a Lite version for up to 1,000 records at a cost of about $1,000 per year (to integrate email and other functions, cost is closer to $2,000), and Bloomerang (which provides the same feature set regardless of how many records you have), with fees ranging from about $1,200 per year for up to 1,000 records, or $6,000 per year for up to 25,001 to 40,000 records.

Sumac offers a free basic CRM for up to 500 records and charges just $240 a year for 500 to 1,000 records, but costs go up to about $2,200 a year for a "Gold" plan that incorporates most of the features described here. Network for Good offers a slightly more comprehensive product, which includes access to coaching, at a comparable price of about $2,100 per year.

In the simple-but-powerful category with DonorPerfect, Bloomerang, Sumac, and Network for Good, Blackbaud offers eTapestry for about $1,500 a year for up to 1,000 records. However, Blackbaud is best known for tools that serve mid-sized or larger organizations and those that derive a significant share of income from individuals.

Blackbaud's products include Raiser's Edge (and the Cloud-based version, Raiser's Edge NXT), Blackbaud CRM, and Luminate CRM. Similar to the all-in-one CRMs described above, Blackbaud is something of a world unto itself with its own knowledge base and helper systems, such as its Everyday Hero crowdfunding system and TargetAnalytics integrated wealth screening. Another product targeted at larger "enterprise-class," Abila Fundraising 50, similarly offers a range of complementary products.

# Buying Additional Data about Your Donors

Your CRM is only as good as the data that's in it. Errors can creep in in many ways—user-generated typos in forms they fill out online, duplicate records, and imports that result in subtly mismatched fields are just a few. You may have email but not postal addresses for some people in your database, and vice versa for others.

Data cleaning and append services—often offered by the same vendors—ensure you continue to deliver to the people you have a relationship list and can help you connect with others. Typically, you upload data to a vendor who can electronically clean-up the list, check its accuracy, and provide additional data for those on the list.

It's easy to buy nuts-and-bolts information like up-to-date email addresses, physical addresses, phone numbers, or geographic coordinates (for mapping purposes) from services like FreshAddress, Melissa Data, TowerData, or True Givers. This can be a good way to get information to jumpstart an email or phone outreach strategy.

Expect the vendor to match just a fraction of your list (perhaps 10-25 percent). Prices vary widely depending on the information you're looking for and the size of your list, but might range from $0.01 to $0.75 per matched name.

The other area of appended data is social. MailChimp has an add-on that will grab your most prolific folks on social media platforms, as will Attentive.ly (purchased in December 2016 by Blackbaud).

You can also buy more detailed demographic data to help assess donors' income levels, often called "wealth screening data." Using public data such as political contributions, home ownership and assessed property values, as well as aggregated donation data from many other sources, wealth-screening firms can offer insight into how much your donors are investing in philanthropy as well as even suggesting other like-minded individuals who might be open to investing in organizations like yours.

Many of these tools integrate directly with the all-in-one and donor management CRMs described here. Wealth screening services like DonorSearch, Blackbaud's TargetAnalytics, DonorTrends, LexisNexis for Development Professionals, WealthEngine, iWave's Prospect Research Online, and DonorScape offer multiple types of data. Many can provide information on either a single donor (through a web interface) or an entire list.

# Volunteer Management Software

Volunteers help with everything from stuffing envelopes to providing specialized legal and other skills to serving on boards. You can choose among many platforms, plug-ins, and back-end management programs to help attract and manage your volunteers.

VolunteerMatch and All for Good are examples of platforms that focus on connecting nonprofits and volunteers. Both offer basic opportunity-marketing and management services free and provide additional functions for a cost. At VolunteerMatch, an annual $99 fee lets volunteers donate via the platform. For $100, All for Good can create a "Project Plugin" to add to your website that displays only your organization's volunteer opportunities. Both services are nonprofits themselves, and have large user communities. "Pro bono" sites connect nonprofits in need of specialized help, such as legal and marketing and communications, with qualified volunteers. Taproot Foundation, CatchaFire, and LinkedIn for Good Volunteering are examples.

Several software companies have served the market of nonprofits managing significant numbers of volunteers for years. Examples include Samaritan Technologies (free to $4,500 per year for standard editions), Volgistics (from $108 to more than $1,000 per year, depending on how many volunteers you will handle), and VolunteerHub ($829 to $2,989 per year for standard versions, depending on number of volunteers tracked and functions used). These programs offer services to schedule and track volunteer hours.

A simpler set of volunteer management tools help with events. They focus on scheduling volunteers, such as to staff a nonprofit or event, and subdividing responsibilities by task, such as setting up multiple booths at a fair. Systems like SignUp.com (formerly VolunteerSpot) and SignUpGenius help you create a list of shifts and responsibilities, upload a list of potential volunteers, message them, and let your volunteers sign up for the shifts and responsibilities of their choice. The system can be set to send your volunteers reminders, too. Both range in price from a free basic package to about $600 per year.

# Specialized Constituent Management Systems

With all these CRM choices, why would you want to get more specialized? A lot of reasons, as it turns out. All CRMs described here can be tailored to your work and organization to some extent. For example, you might choose to add custom fields to capture program-specific data or unique relationships (members, say) that are common in your work. But beyond these modest customizations are some more specialized systems. We profile some types in this section, but there are more—it's worth investigating what constituent tracking software is targeted at your specific sector or need.

### Attendance Tracking Systems

Tracking attendance and time can have applications for **HR and Office Management** systems such as payroll and HR, grant reporting, and **Program Evaluation**.

Unlike other databases, attendance tracking systems often include a hardware component. Readers of Radio-Frequency Identification (RFID)-tagged badges or key fobs are one such relatively affordable solution. Time clocks and biometric systems that scan employee fingerprints are other options; kiosks that employees, volunteers, or visitors to your office can use to sign in are another example.

For larger organizations, Advance Systems' Mitrefinch Time & Attendance System provides a range of capabilities. Lightwork Software integrates with existing HR management systems, exporting its data to Abra, Sage, and Great Plains accounting software. Payroll companies such as ADP and Paychex offer software to help.

## Custom Databases

You'll need a custom database if your organization has a specialized set of fields you need to track that exceeds the amount of customization you can easily do in an off the shelf CRM. Organizations often want to add a way to input and review data via their website or on a mobile device, known as a database app.

Custom database software has evolved to serve these functions, whether for constituent relations management or another purpose. A deep list of news media contacts or an organizational membership database are examples of databases that push the boundaries of traditional CRM products.

Choices of custom databases are available at relatively low cost via TechSoup. Products available include Alpha Five from Alpha Software, which is primarily oriented toward developing web apps. Also available is Filemaker, which, although an Apple product, runs on both Windows and Mac OS. Microsoft Access is another option (offered as part of Office 365 Professional subscription).

## Legal Case Management Software

For legal aid organizations, case management systems track the information that you need to work with a client: contact information, age, address, job history, family relationships, and more. They can also track services you've provided in the past, notes on cases, communications between staff and clients, forms, and other documents. Most case management systems also include reporting features.

Systems that offer more basic features tend to be installed software rather than Cloud-based. ILSForms is a low-cost option that a small office can operate for $500 to $700 per year (does not include consulting fees, maintenance, or training costs). ImmForms 6.8 from Thomson Reuters is a slightly more expensive option at $1,000 to $2,000 per year.

More advanced systems might include online intake, client portals, robust document management, workflow features such as scheduling and deadline reminders, visa bulleting tracking, e-filing, pro bono management, and financial recordkeeping. These systems are hosted by vendors, and users access them through browsers on their computers and, for some systems, their mobile devices.

Some areas of legal aid services organizations will need a system designed specifically for those specialties, such as immigration law, but most legal case management systems can be adapted for different types of services. If your organization works with a diverse group of clients or provides a wide range of legal services you might consider a general legal aid case management system, for example, such as Kemps Case Works, Legal Files, or Pika. The downside is that these systems may not allow you to produce specialized forms, which is an important efficiency for many legal services providers.

When considering a case management system, you'll want the flexibility to configure your case data and to easily create forms, but be careful not to fall into the rabbit hole of customization. You can burn a lot of time and budget tinkering with a system to make it adapt to you rather than adapting to the system. Online intake and document assembly are two of the biggest of these kinds of risks. It's wise to consider systems that have already invested a lot in these areas.

Less common but highly useful features to consider include the ability to manage pro bono volunteers, create automated reminders for upcoming client milestones or court deadlines, and client portals that allow your clients to access specific case information. Cloud-based systems are increasingly common and are recommended for organizations that lack IT support or are not confident in their ability to secure data on in-house servers.

Those looking to integrate client information with data about other constituents (such as donors) or to support complex and unusual processes should also consider *Constituent Relationship Management* (CRM) systems. These systems tend to be very flexible but provide less specialized client-tracking functionality out of the box. For instance, some organizations might adapt Salesforce for use in case management, or use CiviCRM's CiviCase component.

*It's a fact of life: as a nonprofit, you need to raise money. Whether you're raising it from individual donors or grants or through galas or conferences, software can help you raise money and lower your stress level at the same time.*

*All of the software names highlighted within the text are covered in more detail in this guide. Most of them are included in this section, and follow immediately after the descriptions.*

## FUNDRAISING AND EVENTS

# Raising Money and Managing Events

**Whether your primary funding sources are individual donors, grants, or events, fundraising is important to almost every nonprofit—and there's a lot of software that can help.**

### Strongly consider...

Before you even consider the fundraising and event tools in this section, we recommend that you're committed to a database system, which are covered in our Constituent Management section.

Most fundraising programs can benefit from an online component. To get started, you'll need **Online Donations** software that lets you accept credit cards online, either for one-time donations or on a recurring basis.

### Keeping ahead of the curve...

Enlisting staff, volunteers, or close friends to help fundraise is common practice. Several tools are available to help manage the logistics involved. **Crowdfunding** lets you raise money through large groups of people, while **Peer-to-Peer Fundraising** lets supporters create their own online-fundraising web pages, and can support your organization in managing this kind of distributed campaign.

If your fundraising efforts include events, **Event Registration** software can help you keep track of the people coming to your event, take payments online, and record details about attendees such as meal preferences and workshop signups.

Managing the logistics of a complex event can be daunting. Dedicated **Event and Auction Management** software can provide sophisticated support for tracking schedules, budgets, auctions, sponsors, and all the other details that go into planning.

*Online Auctions* can add an interesting online component to your event, as well.

If you're an arts organization and you host performances with assigned seats—such as theater events—you may want a *Ticketing* system, especially one that integrates with your donor or constituent database.

## On the cutting edge...

Your *Constituent Relationship Management (CRM)* software may enable integrations that allow you to access all of these tools in one place.

# Crowdfunding and Peer-to-Peer Fundraising

Crowdfunding and Peer-to-Peer Fundraising use similar processes to raise money, and the tools designed to help share many of the same features.

Crowdfunding raises money through contributions from a large group of people. Peer-to-Peer Fundraising, also called online distributed fundraising, group fundraising, team fundraising, friend-to-friend fundraising, or widget fundraising, involves recruiting supporters—including staff members, core volunteers, or passionate advocates—to fundraise on an organization's behalf. This technique can result in a lot of new donors, but since they're often personally connected to the individual fundraiser rather than the organization, they may not be as likely to participate in other programs or to give multiple times.

In both types of campaigns, users post projects along with their goals and a timeframe, and the community pledges money. Typically the projects include a brief promotional video and offer something in the way of rewards or thank-you gifts to donors, depending on how much is pledged.

One major benefit of Crowdfunding platforms is that they provide nonprofits access to a homegrown network of people interested in supporting compelling projects—frequently to organizations with which they have no established relationship.

The two best known platforms are Kickstarter and Indigogo. Unlike Kickstarter, which is the most popular crowdfunding tool overall, Indiegogo allows charities to raise money on its platform, which makes it the most popular crowdfunding platform for nonprofits.

Indiegogo allows you to set a fixed or flexible fundraising goal. What's the difference? In a fixed goal campaign, you must exceed your goal in order to receive the funds, but a flexible goal campaign allows you to keep the funds you raise whether or not you reach your goal. Indiegogo charges a 5 percent platform fee either way, along with a credit card processing fee. The platform uses FirstGiving for contributions to registered 501(c)3 nonprofits, which also provides tax deductible receipts to your campaign supporters via email.

Many Peer-to-Peer Fundraising tools—including FirstGiving, Razoo, CrowdRise, and Classy—let potential fundraisers create their own personalized fundraising pages, and all offer a version that's free except for a percentage of donations. Pages can have customized pictures, text, and easy links to give money online, and can stand alone or be associated with particular events, like walk-a-thons. Individual fundraisers direct their own friends and family to their pages and take donations. Causes is a unique social network built around the peer-to-peer fundraising model, enabling supporters to engage more directly with the campaigns they support.

More sophisticated tools also let organizational staff members easily oversee a campaign's progress or organize fundraisers into teams. Blackbaud's Friends Asking Friends and TeamRaiser or the tools offered by Artez Interactive, for instance, provide more sophisticated—though considerably more expensive— organizational support in a standalone package. A number of **All-in-One CRMs** and **Donor Management Systems** also offer some of this functionality.

Crowdfunding and Peer-to-Peer Fundraising campaigns take more than just the right tool, however. Each requires planning, experience, and staff time to create and run a successful fundraising campaign. Both start with a strong community of supporters motivated and excited about helping your organization raise money. It's not enough to just pick a tool and turn your supporters loose with it—you'll need to train them to use the tool and to be effective fundraisers, and support them throughout the process with helpful tips, success stories, or inspirational quotes. And once your campaign has ended, you need to recognize them for all the work they've put in to make it a success.

# Event and Auction Management

Planning events and auctions involves a lot of logistics, including sponsors, budgets, schedules, seating, and facilities. Some *Donor Management Systems*, *CRMs*, membership management software, and even some *Volunteer Management Software* can help with these tasks. It's useful to have this functionality integrated with other constituent management functions, but the features these tools provide is rarely as sophisticated as in dedicated packages.

Good event registration software often not only helps your attendees register online, but has event-management capabilities as well—for instance, to assign seats or manage schedules and rooms. Very high-end tools like Lanyon and etouches provide both *Event Registration* and sophisticated management capabilities with functionality designed to help you to track complex logistics, speakers, rooms, and budgets for huge events and conferences. It's very convenient to have management functionality combined with your online registration tool.

Live and silent auctions often require tracking particularly complex sets of information, including the items for sale, their fair market value, buyers, and selling prices, plus the need to generate bills and receipts on-site within minutes of each sale. Dedicated auction management packages like Greater Giving's Auctionpay and ReadySetAuction provide functionality tailored specifically to these needs. Additional tools can be found in the *Online Auction* section.

## Event Management or Event Planner?

No software will plan all the details for you—if your logistics become more complicated than what you can easily manage using mid-priced Event Registration software or a spreadsheet, consider hiring an experienced event planner.

# Event Registration

With the right software, you can accept RSVPs, manage attendee information, and accept payment, all online. For simple RSVP-only needs that don't require the exchange of money, such free tools as Evite, Eventbrite, and Paperless Post, or even SurveyMonkey or Google Forms, provide the ability to understand who's coming, with a little flexibility in how your registration form looks and what information it collects. The Facebook Events application is another interesting option if most of your constituents use the social networking site. Or, you could set up a form on your own *Website*—a straightforward process using a module available with your website **Content Management System**.

If you need to collect registration fees, PayPal, Eventbrite, or Brown Paper Tickets can provide basic functionality with minimal fees. A number of packages, including Click & Pledge, Greater Giving, Formstack, Qgiv, and MemberClicks, support a variety of other transaction types in addition to events like donations and online-store sales. These tools all charge fees that generally come to about 2.5 percent to 4 percent of the transaction, and some also have recurring fees ranging from $20 to $500 per month.

More sophisticated tools like RegOnline and Cvent add support for such additional features as customizing registration page look-and-feel, multi-track conference registration, complex discounts, name-tag generation, sophisticated reporting, and more. Typical pricing includes transaction fees between 2.5 percent and 4 percent of the fee plus $1-4 per registration.

A number of **Donor Management Systems**, **CRMs**, and website **Content Management Systems** support event registration. This is also a common function in membership management systems.

# Foundation Grants Research

At its most basic, **Foundation Grants Research** or grant prospect research consists of two major practices: researching various foundations' grant cycles and giving histories, and managing your organization's applications for each foundation. The former is essentially an exercise in web research—identifying a list of foundations that might give to your organization and locating them online to identify the types of organizations they've funded in the past, and with what size grants. The latter is a matter of tracking and managing data.

When looking up giving histories and grant cycles, websites like the Foundation Center's Foundation Directory Online or GrantStation allow you to search very detailed records of foundations by a variety of criteria, including past grants, focus areas, and giving interests. You can also look to your state nonprofit association, as many will offer access to these tools at a substantial discount. Alternatively, many regional or local philanthropy centers offer access as a benefit of membership, or free on-location in their "grant-research libraries."

In addition, regional associations of grantmakers can be valuable sources. Most grantmaker associations, also known as philanthropy centers, will house a publicly available list of foundations specific to a geographic area. Some are print-only, but a number offer online databases as well. You can also find associations of grantmakers centered around a mission area, such as Media Impact Funders, or on other criteria such as geography, like the Southern California Grantmakers. Searching member lists for these associations may help identify potential grant prospects.

Federal grants are another key source of funding for many organizations. While you won't find these grants in private and corporate foundation databases, you can search for United States federal grant opportunities at Grants.gov. State and local grant listings can most often be found on your municipality's website. A basic web search is also a great way to find out what grants nonprofits similar to your own have received that yours may qualify for.

Many nonprofits also list foundation funders on their websites or in annual reports.

Once you've identified a list of foundations, you'll need to determine their giving histories and grant capacities. While the Foundation Center dataset will have a lot of this information, you may also need to search in other places to find everything you're looking for. Sites like GuideStar let you search a database that contains IRS Form 990s—the form the government uses to track financial information about organizations—from more than 1.8 million nonprofits, including foundations. You'll have to know how to read a 990, however. A number of helpful sites can show you how, including the Nonprofit Coordinating Committee of New York's web page.

As you collect this information, you'll need a place to store it. Smaller organizations can usually accomplish this through a spreadsheet, like in Microsoft Excel or Google Drive, just creating columns to track foundation information, web links, interests, and due dates for RFPs and proposals. Most **Donor Management Systems** will also let you manage your list of foundations just like any other prospect, as well as track RFP and proposal dates, the status of your proposals, and your proposal workflows. If you have very detailed needs around managing grant application processes and requirements, consider a standalone system like Abila Grant Management, Altum, or IT Works.

In addition, it can be useful to supplement your deadline- and submission-date records with **Email and Calendar Software** or task-management software that can function as a to-do list to ensure you don't miss any deadlines or get caught by surprise.

# Online Auction

Holding an Online Auction can be an interesting fundraising technique for a nonprofit, particularly as an adjunct to a live event. They work similarly to traditional auctions—potential bidders browse web pages to see pictures, descriptions, and suggested prices for auction items, and bid by entering the amount they're willing to pay. When the auction ends, the highest bidder wins the item.

eBay for Charity allows nonprofits to conduct auctions on the popular online auction site eBay and receive reduced Paypal processing fees. Listings are available to anyone who searches eBay, which lets you take advantage of the huge audience, but makes it difficult to create a special event feel or to build community.

Other comprehensive auction tools, such as ReadySetAuction, CharityBuzz, and BiddingForGood, are specifically designed for a more "special event" auction feel. They're more expensive than a platform like eBay for Charity, but offer more customized auction homepages and the ability to add sponsor logos, among other features. With these tools, you'll need to drive your own constituents to purchase the items (as opposed to relying on an existing audience, like eBay's), but they are designed to help you create a community feel around your auction.

## Combining On- and Offline Auctions

Conducting an event that involves both on- and offline auctions can expand your audience. ReadySetAuction and Greater Giving (with online auction functionality powered by BiddingForGood) both provide functionality to integrate the two.

# Online Donations

Online Donation software allows you to easily accept credit card payments over the web. Almost all of these tools work the same way—a "donate" button on your website links to a donation form where donors enter contact and credit card information. The tool verifies and charges the credit card securely, and makes sure the money reaches your organization. Online reporting tools let you see what's been donated and export the donation information to other databases.

Many standalone tools can take online donations—in fact, many can also help with a number of different types of online payments, such as membership or registration fees. Increasingly, many organizations use *Donor Management Systems* or *CRMs* that support both online donations and the management of donor data.

Whether as standalone tools or as part of a more comprehensive product, online donation tools typically charge a monthly fee plus a payment processing fee. Such tools as Click & Pledge, Network for Good, Greater Giving, GiftTool, or Qgiv are feature-rich options, offering more support for different types of gifts, customized donation forms, and various campaign types. PayPal offers a simple Online Donation processing option for nonprofits, but does not provide the added features available in other tools. You can also use an online form builder tool like Wufoo to set up an Online Donation form, if you're up for the configuration process.

If you're interested in tools that enable supporters to accept donations for you on their own sites, consider *Peer-to-Peer Fundraising* tools.

# Ticketing

Ticketing is a complicated area, with the potential of a lot of advanced requirements. Even small organizations sometimes hold events to which they want to sell tickets. And pretty much any organization would want to consider patrons who have bought tickets as donors. But when you put these two together, it can make for a complex and often expensive system.

At the lowest end of the spectrum, you can use the same vendors you would use for paid event registration. Vendors in this space, such as TicketLeap, Eventbrite, or Brown Paper Tickets, allow you to sell tickets online, including different levels of tickets to the same event—for example, VIP or mezzanine tickets. (For a broader overview, see *Event Registration.*)

If you just need to sell reserved seats for your venue, and it's not critical to you to easily pull your event patrons into your fundraising process, there's a lot of small online software packages that will allow you to sell tickets. These sites—like OvationTix, Tix, Arts People, and Vendini—allow you to sell tickets, often including box office functionality to print tickets to box office printers. Center Stage Software's WinTix/WebTix provide both installed box office software and online solutions.

Even for a small organization, it can be very useful to have a single system that allows you to track not just tickets but also the full fundraising process. Easy-Ware's Total Info and Arts Management Systems' Theatre Manager provide affordable installed software with integrated online ticketing functionality.

In addition, it can be valuable for organizations of all sizes to track their patrons with their other constituents in a *CRM*. For small- to medium-sized organizations, PatronManager is built on top of the Salesforce platform and includes reserved seating ticketing and box-office functionality. Choice Ticketing Systems and AudienceView provide more sophisticated functionality and provide some constituent tracking functionality for medium-sized and larger organizations. Spectra also provides ticketing and fundraising functionality for larger organizations, and Blackbaud's Altru also provides box office functionality, with some ability to integrate into The Raiser's Edge fundraising platform.

For the arts management world, Tessitura provides functionality that integrates complex box office, online ticketing, and fundraising. It's a complex system that's more appropriate for organizations with multimillion dollar budgets than for small ones, and requires considerable customization, training, and staff time to use.

If you have complex box office needs but don't need a lot of fundraising functionality, you may find that the ticketing systems used primarily by stadiums and commercial venues work well for your needs. These systems, like ProVenue by Tickets.com or Ticketmaster, tend to not offer much sophisticated functionality to track donor interactions, pledges, or a donation made with an event payment, but instead focus on complex online and offline box office functionality needed by venue.

*To build your base, you'll need to reach out to find new supporters. There are a number of ways to get your story out there to make it easier for people find you. From software to help with Broadcast Texting to Graphics and Multimedia, there are tools to help any nonprofit.*

*All of the software names highlighted within the text are covered in more detail in this guide. Most of them are included in this section, and follow immediately after the descriptions.*

**COMMUNICATIONS**

# Communications

Whether you want to build your base, strengthen relationships, fundraise, learn from your constituents, or create community dialogue, you'll need a mix of communications tools and strategies to help you reach your goals. A successful communications strategy requires a reliable **CRM** and an understanding of the various communications channels available to your organization.

Strongly consider...

A communications strategy document and ongoing communications schedule are essential for smart *Communications Planning*.

Once you have a plan in place you can send *Broadcast Email* and develop *Website Content* that provide valuable information or encourage people to take action. Make sure you adopt a user-friendly website *Content Management System (CMS)* for your website that helps with *Search Engine Optimization (SEO)*. Also make sure that yours is a *Mobile-Friendly Website*. For most websites, more than half the visitors are arriving via mobile devices—if your site isn't easy to read and navigate on a mobile device, you're going to lose visitors and drop down the search rankings.

Your *Social Media* presence can help you amplify your messages and reach new people. Images are more important than ever on social media. Nonprofits that have in-house *Page Layout* and *Photo Editing* skills can create exciting content that will get noticed online. *Charts, Diagrams, and Infographics* can also increase engagement on social media.

## Keeping ahead of the curve...

Nonprofits that run *eAdvocacy* software have a dynamic platform that enables them to mobilize their follows to effect change. Some of these tools allow you to create **Landing Pages and Microsites** that enable you to immerse the audience in a topic and focus your call to action. Microsites often rely on multimedia content so to create an exciting site, you might want to bring on the tools and skills needed for **Multimedia Editing**.

Some nonprofits lead **Webinars** using specialized platforms that enable dozens of people to join an online event that includes video and audio. These are typically nonprofits with members who are seeking information or nonprofits that have education as a central part of their missions.

## On the cutting edge...

Unless your organization has a dedicated following that needs information or services quickly and on the go, you probably don't need a mobile app. However, many nonprofits are discovering that they can reach mobile users through **Broadcast Texting**. In addition to **Facebook**, **LinkedIn**, and **Twitter**, **Other Social Media Platforms** can be useful for reaching out to new people, but learn the culture around each platform first and make sure that the platform you choose matches the people you're targeting. Some nonprofits even use **Online Advertising**, but that can be expensive and difficult to convert into meaningful action.

# Communications Planning

Communications professionals like to say, "If you don't have a plan, you're planning to fail." It can be tempting to write and send a **Broadcast Email** or post to **Twitter** when inspiration strikes, or when you have a spare minute, but the truth is that you're likely missing opportunities and failing to engage your audience.

Communications Planning helps you draw on content and insights from across your organization. It helps you build momentum for a cause and make sure you're part of the conversation during major events. Most importantly, it gives you the time and space you need to develop content your followers value, and it allows you to make your communications channels an extension of your mission.

The first step in developing a communications plan is creating a calendar. Many organizations use the **Email and Calendaring** software already in place—Google Calendar and Microsoft Outlook are two of the most popular. These platforms allow you to schedule communications activities and share them with teammates, who will then see the tasks or deliverables on their own calendars. And when you view the communications calendar on its own, you're able to see the sequence of communications and better understand whether you're getting the timing right.

Some communications professionals prefer to use spreadsheets to map out their communications on different channels. Spreadsheets have the advantage of flexibility—you can include a lot of information and use formulas that dynamically update other cells when you make an edit. If you're developing a large campaign with many stages, you can also use a spreadsheet to create a Gantt chart that shows the sequence and timing of each stage.

**Project Management** tools are also useful for Communications Planning, especially campaigns. They can help with timing, budgeting, and managing the people who need to contribute. There are many free project management tools that might be useful including Zoho Projects, Teamwork Projects, Wrike, Volerro, and ProofHub.

Keep in mind, Communications Planning requires a lot more than making a schedule and managing people. A good plan also includes clear goals and a keen understanding of which audiences will help you achieve those goals. Knowing your audience and goals then makes it easier to decide which channels will be most effective. For example, if you want artsy millennial women to participate in an auction, you might choose to post interesting photographs of auction items on Instagram, but focus on event details on *Facebook*, and mostly ignore *Twitter*.

There is software designed to help you make and maintain a communications plan. Smart Chart 3.0 from Spitfire Strategies takes you through the entire communications planning process from goals and objectives to measuring your success. At the end of the process you can print out a chart that documents your planning and charts the stages and timing of your campaign.

It takes thoughtful discipline to develop a plan and carry it out, but it's worth the time and effort and it will get easier as planning becomes part of your organization's culture.

# Crowdsourcing

Crowdsourcing is less a technology than a strategy for enlisting help or answers from the internet. Unlike the similarly named "outsourcing," Crowdsourcing relies on the assistance and expertise of the broader internet audience (the "crowd") rather than a specific individual. The process has a long history, from humble beginnings as a general open-call for help on discussion boards, but in recent years *Social Media* and new technologies have made it easier for organizations to reach and engage broader audiences.

Nonprofits have used Crowdsourcing for everything from marketing and fundraising to volunteerism and activism. It's a great way to enlist help from a wider community knowledgebase and to engage people in your work.

If you're trying to pool collective knowledge on a specific question or subject, the open-call technique still works, whether through social networks (such as *Facebook*, *Twitter*, or *LinkedIn*), *Email Discussion Lists*, or through specific Crowdsourcing platforms such as Ushahidi for collecting geographic data on certain types of incidents.

If you're trying to do a specific task instead of just collecting information or answers, you can also turn to the crowd, where there are many platforms that connect you with experts or volunteers willing to jump into a short project. Some platforms, such as Skills for Change (powered by Sparked) connect you with volunteers for small tasks. You can find volunteers with specific expertise through Taproot+ and Catchafire. If you're willing to pay a little for help, Lingotek can help with translation, 99designs lets your review multiple designs before choosing the right one for you, and Amazon's Mechanical Turk can be used to complete simple tasks or hire research subjects. You might also explore *Crowdfunding* tools such as Kickstarter or Indiegogo, or microfinance sites like Kiva.

Overall, Crowdsourcing will be more effective if you make it very clear what you want the crowd to do, break down your strategic goals into smaller tasks people can help with, engage the crowd and reward participation, and stay positive.

# eAdvocacy

Organizations use a number of online tactics to get constituents involved, including asking them to send emails to decision-makers or politicians, or to take other actions on their behalf. Tools for supporting online actions—often known as eAdvocacy tools—make it easier to encourage and manage such techniques.

Reaching Congressional representatives in bulk can be difficult, as many of them block automated emails. You could ask constituents to find and email representatives for free through Congress.org, but you're likely going to need a system that makes it easy for supporters to take action and for the actions to reach people in power.

Organizations that need a straightforward, low-cost tool that doesn't need to be integrated with a **Donor Management System** or other fundraising technology might consider Voter Voice. For as little as $500 per year you can run letter and email campaigns.

Some organizations use integrated online systems (such as SalsaLab's Salsa and Blackbaud's Luminate) to help supporters draft and send emails that reach the appropriate audience. These systems also include **CRM** and donor management capabilities.

A number of advocacy solutions now offer a variety of outreach and advocacy tools. For organizations that need a short-term tool or only occasionally run advocacy campaigns, Spark Influence and Phone2Action are mobile-friendly platforms that let you reach out to supporters via email and text message, identify influencers and top activists, and mobilize your supporters to contact decision makers via email, phone, and **Social Media**.

NationBuilder is the most well-known platform in this space and boasts integrations with other systems and services, including Phone2Action. Attentive.ly (recently purchased by Blackbaud) is focused on social media advocacy, but also is good at bringing social media data into your CRM. The Action Network is a solid general option and is free for individuals and small groups.

Alternatively you could ask supporters to write letters to the editor of their local papers. You'll also want to provide contact information for media outlets—again, some more-advanced integrated online systems support this.

Services have also popped up that aim to replace email as a form of advocacy. Tools such as POPVOX let your supporters create profiles, upload their voting history, and directly message their representatives. It's unclear how effective these tools are for advocacy, but as usage increases, they may become as valuable as email.

## Advocacy Through Social Media?

Nearly every company and politician is on social media, so there are often opportunities for your supporters to reach out to advocacy targets through *Twitter* or *Facebook*. If many supporters post, it can provide a public demonstration of the support for your cause and can be amplified by people who aren't on your mailing list. *Twitter* is still the most popular social network for politicians in particular—your supporters might be encouraged to TwitterBomb a legislator with hashtagged content to encourage her to support a certain position, for instance. Many eAdvocacy platforms now support social media action.

# Email

The demise of email is greatly exaggerated. According to MailChimp, about 26 percent of nonprofit emails are opened—that's four times more than the average business (6 percent) and many more readers than you're likely to get on *Social Media*, unless you pay for *Online Advertising*.

It's not hard to get started engaging your audience via email. The first step is to collect the email addresses of supporters who want more information.

You can start with your mailing list. Email appending services such as FreshAddress and Datafinder match email addresses to your snail mail subscribers. It's typical to match only about 20 percent of your list. But tread with care when using email addresses acquired through an append. Some constituents might prefer not to get email from you, so make sure it's easy for them to opt out.

Your *Website* can drive sign-ups, especially if you frequently provide useful and timely *Website Content*. The simplest way to do this is by including a link to a form that collects subscriber information and transfers it to your *CRM* or *Broadcast Email* system.

If you hold events, you can include an eNewsletter opt-in button on the *Event Registration* form. If your approach to events is more low tech, put out a sign-up sheet at your next meeting and type the information into your system manually.

Once people are signed up, make sure you're giving them something worth subscribing to. If all of your emails are fundraising pitches, you're going to lose a lot of subscribers. Instead, think about why they want to follow your organization and what kinds of information or resource they will find valuable. Many organizations frequently add useful content to their website or blog. Include summaries and links to that content in a regular newsletter email. This will not only make people want to open your emails, it will also drive traffic to your website.

## Broadcast Email

Broadcast Email software lets you email a group of people all at once—as many as you want. It also helps you create attractive emails (often through graphic templates), manage email address lists, and enable people to subscribe and unsubscribe by themselves. In addition, most tools help you collect email addresses on your website, "mail-merge" information into emails, send messages to particular segments of your mailing list grouped by demographics, and report on how many recipients opened or clicked on each email.

VerticalResponse is a good option in this area, offering nonprofits a sophisticated feature-set and up to 10,000 emails per month for free. For example, if your list has 1,000 addresses on it, then you can send 10 messages to your entire list without paying any fees. Nonprofits that send more emails than 10,000 per month can get 15 percent off their monthly plan.

MailChimp offers a similar deal—up to 2,000 subscribers and 12,000 emails per month for free. Both small businesses and nonprofits can take advantage of this deal. Constant Contact is slightly more expensive, but might be useful for associations. Campaign Monitor, Emma, and iContact are user friendly and offer valuable data and automation tools, but will likely cost more than $1,000 per year. WhatCounts also offers agency services for a more holistic approach to email, but will be expensive over time.

Note that many **Donor Management Systems** and **CRMs** provide some Broadcast Email functionality. While not all can match the advanced features of dedicated software, you may find that your existing software meets your needs. It's also worth setting up an integration between your constituent management solution and your broadcast email provider to get a more holistic picture of your supporters.

# Take Care with Installed Packages

If you email more than a few dozen people at a time, use an offsite vendor that supports mass emails. Tools like Outlook aren't designed to support large-scale mailings, and won't help with the formatting and list-management tasks critical for large lists. When you use them to email hundreds of people, you may reach more Spam filters than inboxes—or worse, your mail server might be blacklisted as Spam, blocking future emails from anyone in your organization.

## Email Discussion Lists

Email discussion lists, often called "listservs" for the original software application that supported them, let people subscribe to topic- or group-oriented discussions. When a subscriber emails a specific automated address, the message is emailed out to all of the list's subscribers. Other subscribers respond, and an ongoing discussion unfolds in inboxes. As email is familiar to a wide audience, email discussion lists can be a straightforward way to encourage people to talk to each other online. Typically, you'd use a discussion list if the conversation or information is public and widely sharable.

Most of these tools let people subscribe or unsubscribe, decide whether to receive messages as they're sent or in digest form, and choose to view messages either via email or in a web interface. Most let users search archives online. Administrators can moderate emails, and more advanced tools let you customize the graphic design of the interface and more easily integrate the email addresses with your other databases.

Both Google Groups and Yahoo! Groups offer basic discussion list functionality for free, but include prominent ads in the email messages. Electric Ember's NPOGroups offers similar functionality without the ads. Groupsite adds a bit more power, but pricing starts at $30 per month. LSoft is a high-end and customizable solution that lets you have a listserv with your own domain name and can incorporate email marketing tools.

# Graphics and Multimedia

A decade ago, graphics and video were a luxury for nonprofits, and often out of their reach. Today, if your organization isn't creating visual content it will likely get ignored, especially during fundraising drives. Online supporters—most notably on **Social Media**—expect your communications to be fun, attractive, and visually interesting.

You're also wise to consider creating **Charts, Diagrams, and Infographics** to share with funders and other supporters. Numbers alone are not good at getting people excited about your progress or motivated to help. Visualizing your data allows you to not only show what you've done but why it matters—and how it fits into the larger story about your work.

To create valuable visual content, you'll need the right tools, especially if staff members will need to teach themselves new skills. Consider these tools and techniques as you work on making your content more visually engaging.

## Charts, Diagrams, and Infographics

There's a reason people say pictures are worth a thousand words. Graphical depictions of data or processes can communicate important information in a single glance. Data visualizations are more approachable and usually more interesting than the raw data. It's even possible that your visualizations—especially when part of an infographic—will be shared widely and that you'll reach a new audience.

Google Sheets provides tools geared toward creating interactive online charts inexpensively, or even for free. However, you won't be able to create charts that are high enough quality for printed publications. Microsoft Excel, SmartDraw, and DeltaGraph provide powerful functionality to create printed charts, and they're all under $200 per year for nonprofits. Tableau offers sophisticated online and offline software geared toward those with more complex needs and a higher budget. For more complicated visualizations, you'll either need statistical analysis tools for more mathematical power, or the advanced abilities of software such as Adobe Illustrator.

If you'd like to create diagrams that are not specifically based on data, Visio, SmartDraw, Gliffy, or OmniGraffle can help by providing predefined shapes and templates for diagrams such as flowcharts, org charts, or user interfaces. It can also be useful to create free-form "mind-maps" using tools such as Mindjet, MindMeister, or FreeMind. These tools help you show the relationships between ideas or information in summary format and between parts and wholes. For example, a mind map can be used to represent the objectives and strategies or components of an organization, program, or campaign.

Prezi also provides an interesting, interactive diagramming and presentation tool that allows you to lay out all your information in one large view, and then zoom in and out to focus on different areas.

In recent years, a handful of web-based infographic creation tools have emerged. Easel.ly, Infogr.am, and Piktochart all provide templates you can use as a starting point to develop your infographic. All three offer a free option, but for more templates and features you will need to spend a few dollars per month for a subscription.

## Collaborating on Diagrams

Want to work together with a group? Gliffy and Google Drive's diagramming tools allow group members to create or update diagrams and flowcharts together in real time over the web. VoiceThread takes on collaboration and commentary in a different way—it allows anyone to make remarks on almost any form of media (including diagrams, charts, drawings, photos, or video) via audio overlay, video popup, or text box.

## Multimedia Editing

Multimedia Editing software gives you the ability to create videos or podcasts with a level of a polish that used to require a lot of expensive hardware. Good editing takes time and some skill, but a number of low-cost, straightforward packages put the tools within reach of any nonprofit. There are a lot of free and low-cost **Mobile Apps** and web-based software that nonprofits can use, too.

With audio packages, you can edit interviews for length, cut "um"s and pauses, and add music or voiceover introductions. Both GarageBand (for Mac) and Audacity are free, solid tools that provide all the functionality you're likely to need. Sophisticated paid tools include Adobe Audition, WavePad (free for non-commercial use), and Acoustica Digital Audio Editor.

Video tools let you cut out footage you don't want, splice different sections together, and overlay graphics and text onto your piece. You might join an interview with a constituent together with scenes of your program participants, and put a title screen and music at the beginning—and with a single click, even upload the video to YouTube or Vimeo.

It is possible to over-edit your video. All of these software options come with special effects that can enhance your video, but don't overuse them—a little goes a long way.

If you want to create a short, simple video, consider apps such as Camera Plus Pro, FiLMiC Pro, and Instagram Video. There are also apps that can help you edit longer videos and many are surprisingly robust. These include PowerDirector, KineMaster, and Movie Edit Touch.

If you need desktop software video editing, Mac users have iMovie (free with the Mac operating system), a great tool for simple movies. The free editing software available for PCs, on the other hand—such as Windows Movie Maker (soon to be discontinued)—can be difficult to work with and often imposes confusing limitations on supported formats. Alternatively, for friendly features very similar to iMovie, consider Adobe Premiere Elements ($27 on TechSoup when bundled with Photoshop Elements), Filmora Video Editor, or VirtualDub.

If you've outgrown the low-cost options, or want to create more robust animations or special effects, Final Cut Pro is a logical stepping stone for Mac users, while Adobe Premiere is a popular option for both Macs and PCs. These products, each just a few hundred dollars, provide a lot of power. If you're skilled or savvy-enough and have high-end video production needs—and a budget to support them—there are numerous video solutions available on the market. Avid Technology makes the most well-known, including the core of its product suite, Media Composer, which costs $1,299 and requires a powerful computer. However, if you think you need such powerful tools, you're likely better off contracting a professional video editor.

Once you've edited your movies, it's easy to get them up onto the web—see *Video Sharing and Streaming* for more information.

## Page Layout

Page Layout software enables you to arrange design elements (such as text, images, colors, and lines) to create newsletters, posters, reports, invitations, and other materials. Appealing layouts require graphic design skill as well as software, but good Page Layout software can make your job easier and put high-quality designs within reach of anyone willing to learn a few basics.

Straightforward projects such as reports or text-heavy posters can be designed in word processing applications (Microsoft Word, Google Drive, or Apache OpenOffice Writer, for example). These tools offer more control over text and images than you might think, but are difficult to use if you want to layer images, incorporate multiple design elements, or work with complex text flow and spacing. These tools also will not allow you to save in formats required for professional printing. Other office productivity tools can be used for layout: Microsoft PowerPoint and other presentation software are more flexible ways to arrange and layer images, and Microsoft Publisher offers even more sophisticated control without requiring the skill needed for more complex tools.

Unlike word processing applications, professional Page Layout software treats pages as a series of distinct elements so you can format, edit, or rearrange them independently. The most popular tool in this category by far is Adobe InDesign. Nearly every professional designer is proficient in InDesign, and it's the preferred tool for most of them. Generally, if you will need to modify an asset in the future or work with a professional designer on your project, it's risky to use anything other than Adobe InDesign, as files will likely need to be redesigned. However, if you need a low-cost option that has sophisticated layout features, you may have success with Scribus (free and open source) or QuarkXPress.

Professional layout software will speed up the layout process enormously over a word processing tool, but the complex menus and features make these packages difficult to learn. In many cases you'll need a solid foundation in graphic design lingo to understand their terminology. Investing in a good book or taking a class will get you on the right track.

If you need to create a quick poster or social media image and don't have Page Layout skills, there are a few online tools that provide numerous templates that allow you to simply swap out text or add images to create your own professional-looking design. Canva is free for up to 10 users and surprisingly flexible. Lucidpress offers numerous print templates and drag-and-drop tools for between $5.95 and $40 per month. In addition, many infographic tools such as Piktochart or Venngage can be used as basic layout tools.

## Photo Editing

Digital cameras have put professional-quality photos within everyone's reach, but once you have a high-resolution image, how do you prepare it for print or the web?

Cropping—trimming an image to remove unwanted items or to isolate the subject—is often needed to make a photo look professional. Color correction—for example, boosting a washed-out photo's color, or removing red eye—can save a less-than-perfect image. And resizing can make an image the appropriate dimensions for the target media and shrink file sizes to make websites load more quickly.

The right Photo Editing software can do all these tasks and many more, but tools in this area usually offer a trade-off between power, usability, and price.

Newer smartphones now take very good pictures and a wide variety of free and paid photo editing apps have emerged. Instagram is a popular tool that lets users apply various filters to photos and upload them to an online profile. Other apps such as Adobe Lightroom Mobile, VSCO, Enlight, and Google Snapseed allow you to make quick, sophisticated edits.

Free tools aimed at the desktop user, such as iPhoto, PhotoScape, and Pixlr, are simpler to use than professional applications. Most provide plenty of functionality for simple image-correcting and cropping and some advanced features such as multiple image layers or the ability to cut a person out of a background and paste them into a new location.

Middle-tier packages such as Adobe Photoshop Elements (available on TechSoup for $27), Google Nik Collection (free) or Paint.NET (free) add additional power, but also additional complexity. Web-based tools such as PicMonkey (ranging from free to $7.99 per month) and SumoPaint (free to $4 per month) offer the ability to get creative with your photos.

Adobe Photoshop is widely recognized as the most powerful Photo Editing software available. Adobe offers Lightroom, Photoshop, and some **Mobile Apps** as part of its Creative Cloud Photography package for $9.99 per month. Photoshop rivals include Serif Affinity Photo, ON1 Photo 10, and Corel Paintshop Pro X8. GIMP, a free open source program, offers many of the same features as Photoshop, but in a somewhat less-intuitive interface. Even tech-savvy users will benefit from some form of instruction on these packages, whether through a book, a class, or one-on-one time with a graphic designer.

# Mobile

For many people, a mobile device—a smartphone, tablet, or smartwatch—is the primary way they access online content. In fact, according to a 2016 study from the Joan Ganz Cooney Center at Sesame Workshop, 41 percent of Hispanic families can only access the internet via a mobile device.

A large percentage of your clients and supporters don't have a computer, but they do have an internet-connected phone. For your organization to deliver its biggest impact, you'll need to integrate mobile strategies into your program delivery, advocacy, communications, and fundraising.

To adapt to an increasingly mobile world, you'll have to understand the people you're trying to reach and how they use their mobile devices. Is your **Website** easy to read on a mobile device? Do your constituents prefer to communicate by text message? Will a simple **Mobile App** solve a persistent problem? Thinking through why people come to you and streamlining your communications and interactions with mobile device users can make your information and services easier to access and even make your organization more efficient.

## Broadcast Texting

Most cell phone carriers now include text messaging with their service, making it increasingly desirable for nonprofits to reach out via cell phone text messages (also known as SMS or Short Message Service). Make sure to get users' permission before sending them texts, however—the law requires it, and the people who have to pay to receive texts will not appreciate the unsolicited message.

Texting doesn't have to be expensive. Google Voice lets you send text messages to up to five recipients from the Google Voice website, in the Google Voice app, through your SMS messaging app or through Hangouts. BulkSMS allows you to pay per text message, at about $0.04 per text.

There are a number of online fundraising tools that function a lot like **Broadcast Email** tools. They usually include the ability to send text messages to targeted sets of subscribers, collect data, provide interactive responses, and manage subscriptions and unsubscribe requests. MobileCause provides a full suite of fundraising outreach tools including email, texting, and phone banking. Mobile Accord/mGive focuses primarily on mobile giving. Mobile Commons allows integration with many **CRMs**. Check with the vendors for nonprofit pricing. Many **eAdvocacy** tools also now include texting outreach.

Most of these services also support mobile giving, which allows subscribers to donate $5, $10, or $25 added directly to their phone bills. It's not currently possible for them to donate more than $25 at a time, and you'll need to wait until the subscriber pays their phone bill—sometimes as long as several months—to receive the money.

Texting's potential isn't limited to mobile giving, however. Texting works best as a two-way channel. This could be automated using a "branching" set of responses. For example, a supporter who texts "HELP" to a particular number could get information back asking if they're able to volunteer at an upcoming event. If they answer "YES," they get information about the event in response; if they answer "NO," they might be offered another possible opportunity. A system can also automatically pull responses or information from a database.

In another example, the tool Ushahidi lets you display texts asking for help on a map—in one instance, it was used after the Haiti earthquake to plot areas of greatest need and help locate people who needed to be rescued. See **Maps and Geographical Information Systems** for more information on how geographic visualization can be useful in program evaluation.

Additional low-cost, general Broadcast Texting services include FrontlineSMS, Yakstak, and Textmarks. For a more personalized approach, Hustle emerged during the Bernie Sanders campaign to facilitate one-to-one texting to many people. Essentially, it's like Broadcast Texting, but with real people using real numbers that recipients can text back to start a conversation.

Options such as Clickatell or mBlox offer services that allow programmers to automatically send texts from another application or pull text into a database via an Application Programmer Interface (API). And there are a number of free tools, such as Kannel or Gammu, that allow you to connect your own cell phone to your computer (or if you need to support a larger number of texts, you could connect a GMS modem to your computer instead of a cell phone) to send bulk text messages. This is a free, robust way to send texts, especially in developing countries, but U.S. carriers frown upon the practice and are likely to shut down your account.

## Mobile Apps

An increasing number of your constituents are using smartphones or tablet computers. These devices let them browse the web and download mobile applications, or "apps," for more interactive functionality. The easiest ways to reach people through their phones are **Mobile-Friendly Websites** or **Broadcast Texting**, but if you're looking to provide a specific functionality, you might consider creating an app.

Apps can engage constituents or provide them with useful information. Staff and volunteers could also use them to carry out organizational work. Unlike a mobile website, which can only be accessed with a Wi-Fi or data connection, apps generally don't need internet access to be useful, but your constituents will need internet access to download them in the first place.

The many different smartphone platforms create a challenge because Mobile Apps are platform-dependent—those designed for iPhones won't run on Android, and vice versa. You need to either make educated assumptions about your users and desired audience or build multiple apps for competing platforms.

Free apps still dominate the marketplace, but users have shown a willingness to pay a few dollars for apps that manage to be both useful and cool. And that's the challenge—for an app to be popular, it has to be useful. This sounds like common sense, but there's no shortage of businesses creating apps for brand dissemination that don't add any value.

While there are a few tools that allow non-technical users to create very simple mobile apps—such as Appy Pie or AppMakr—for the most part, creating a Mobile App will require a programmer. This can be expensive and time consuming, so make sure your app meets a fundamental need before developing one.

## Mobile Apps for Attendance

Mobile Apps can be a good option for tracking attendance and participation at large, in-person events and trainings due to the availability and portability of mobile phones. There are two types of apps in this area—those that attendees use to "check-in" on their own phones, and those that your staff members can use to take attendance.

Mobile Apps in the first group, like SmartConnect (formerly Geniemobile) from Genie Connect and QuickMobile, allow attendees to check-in, create schedules, and even share notes with other attendees. The downside of these apps, however, is that they require attendees to download them beforehand.

Mobile Apps in the second group, like Event Check-In for Constant Contact, let your staff take attendance on their own phones, either by "checking-in" registrants or scanning QR codes. You could also look at apps designed for teachers, like The Attendance App, Attendance IQ, Attendance, Attendance Tracker, or Meeting Attendance.

## Mobile-Friendly Websites

Most U.S.-based web searches take place on mobile devices, according to Google. To make sure your **Website** is easy to read and use on a wide range of smaller devices, you need to design it to be responsive to the screen dimensions of the user's device. Fortunately, designing a website to work well on mobile devices isn't as hard as it might sound.

Smartphones will show almost any website, but some look better than others. If you use a phone to look at a website that was designed for desktop screens, you will usually see either the upper left hand corner of the website (because that's all that will fit on the small and vertically-oriented phone screen) or a tiny and hard-to-read version of the entire page, shrunk to fit the phone screen.

If it's simply not possible to redesign your website anytime soon, consider tweaks to make it easier to view the site on a phone—for example, place key navigation elements in the top left corner to allow mobile users to browse without scrolling horizontally. You also might make navigation elements big enough to be visible even if the whole website shows up in a tiny version.

However, the more scalable and strategic way to support mobile devices is to build your graphic design using what's called "responsive design." This is a way of coding your website template to rearrange and scale the text, images, and layout of a web page depending on the size of the browser reading it. Your responsive website might appear with three columns and large images on a computer, but only one column and small images on a smartphone. In order to support responsive design, you'll likely need to update the way your graphical design is implemented.

# Social Media

The average *Facebook* user checks her feed more than a dozen times a day and spends more than an hour reading, "liking," and commenting on posts. The promise of Social Media is the opportunity to reach out to people where they like to spend their time. But you're also part of the noise of daily life. When someone views your organization's *Facebook* or *Twitter* posts they're seeing them within a stream that also includes baby photos, vacation updates, news stories, and random observations and musings. How do you get noticed when there are so many other posts to grab their attention?

First, it's important to be real. Social Media is unforgiving to people or organizations that are putting on an act. Don't try to piggyback on a trend or be part of a conversation that isn't a natural fit. Show your organization and its people as they really are. Tell stories that humanize you and your work and share information that's relevant to what you do. Your followers want to get to know you and to see inside your world. Show them what it takes to make your organization run and show them how it has changed the lives of real people. That authenticity will keep your followers in touch and may spur them to act when you need help.

However, don't focus too much on promoting yourself. If donors are your primary audience, show them that you care about the issues. Share content from organizations with similar values and objectives or news that they will likely care about. The most successful nonprofit Social Media pages provide information or create experiences that are meaningful to people beyond the work you do.

Social Media posts that get noticed usually include images or multimedia and not many words. People scroll through their feeds fast—you get less than a second to make an impression. Wordy posts with no visuals are almost certain to get passed by.

Most importantly, have fun. That's why your followers are there. Don't be afraid to be quirky once in a while or do something a little silly (as long as it is not inappropriate or wildly off brand).

# Facebook

Facebook has nearly two billion monthly users worldwide, making it the largest social network in the world.

To get started on Facebook you need to create a page for your organization and invite current or potential constituents to "like" your page. (Note that all admins of your organization's page will need a personal Facebook account.) Once your page is set up with basic information about your organization, you can then promote events, post information, host discussions, share articles or photographs, and solicit donations.

A few years ago, nonprofits could post content and be reasonably confident that a high percentage of their followers would see it. Not anymore. Unless you pay to "boost" your content, it's not uncommon for a post to reach only 3 or 4 percent of your total audience. Fortunately, boosting and *Online Advertising* through Facebook is relatively inexpensive. For $10 you can reach up to 1,000 people. Facebook Ads, the advertising management app built into the platform, allows you to target followers and non-followers by numerous categories including gender, location, income level, and interests—including other specific pages they follow.

Facebook offers closed groups, which can be a useful way to comply with privacy or confidentiality guidelines, or simply provide a safe place for your constituents to interact with each other. It also offers a number of apps that can expand the functionality of your page. For instance, you can use Lead Forms to build an email capture page, and marketing services such as Tradable Bits let you drive campaigns to your supporters. Facebook has also released new features to help nonprofits fundraise. You can now create a donate button for your page and followers can create *Crowdfunding* campaigns directly within the Facebook platform.

With Facebook, as with all Social Media, results will vary widely among different organizations. You should consider your audience and your mission, as well as how much time you'll be able to invest, when creating a social media strategy. Also, don't overlook the importance of visual content. Meaningful images can drive engagement and help your content reach a larger audience.

# LinkedIn

With more than 450 million members, LinkedIn is an interesting site that falls somewhere between social networking for the general public and a niche social networking site for professionals. A LinkedIn profile is essentially a resume. People describe their work history and skills, and can include educational background, references, associations, and more. Each person can "connect" their profile to other people they know—when you make a "connection," you see their profiles and the people their connected to on LinkedIn. You'll need to set up a company profile for yourself and your organization first—both are free.

There are two main benefits to using LinkedIn. It's both a useful place to search networks to find potential contacts and new friends and it's a great place to share information and insights with groups of people who work in a similar profession.

As a networking tool, LinkedIn can help you find new donors, staff members, volunteers, or board members, especially those with specific skill sets. You can do this yourself through the platform's search functionality or can pay for LinkedIn Recruiter to search its member database and send you a list of potential applicants. Recruiter Lite is available for free, but with limited search capability. Nonprofits may be able to get a discount on the full Recruiter service.

LinkedIn is likely to be of particular interest to groups whose mission is to support people in their jobs or who have a focus on careers. As an organization, you can create a LinkedIn group for people to join. Members can hold discussions, post resources of interest, or create a job board. Group members also receive digest emails of the discussions and postings. As an organization, you can select what content is included in these emails.

LinkedIn also includes professional development training, job postings, and a place to post presentation decks called Slideshare.

# Twitter

Twitter is a popular social networking platform that lets you create a profile for your organization and send out a stream of short messages called "tweets"—updates about what you're doing, conversation starters, requests for help, or links to resources of interest. Tweets are limited to 140 characters or less, but that no longer includes screen names and media attachments.

You can use the Twitter.com website to manage your account, but many people instead choose to tweet via applications such as TweetDeck or HootSuite because they allow more sophisticated management of incoming and outgoing tweets—both on smartphones and computer desktops.

People can choose to "follow" your Twitter account, which is like subscribing to your feed, and if they particularly like one of your tweets they can "retweet" it, or post it again so their own followers see it. It's in retweeting that much of the power of Twitter lies. If you post something interesting that's retweeted exponentially, you can reach a huge number of people very quickly. Just make sure you add a comment to your retweet and add some value—otherwise it's easy to overlook retweets.

You can also use "hashtags" to post a tweet to a certain group. For instance, including the #nptech tag will flag your post as relating to nonprofit technology and make it more likely to be seen by those following #nptech tweets. The # symbol, or hashtag, makes your keyword or phrase easily searchable by others—various sites even track existing hashtags you can search before creating your own.

Many individuals and organizations also use URL shorteners such as Bit.ly, Ow.ly, or TinyURL to make posts look less long and messy. It's also now easy to attach photos, gifs (one click gives you access to a large library), polls, and a location to your tweet.

With all social media, results will vary widely among different organizations. Twitter users tend to be more tech-savvy. You should consider your audience and your mission, as well as how much time you'll be able to invest, before assuming that Twitter makes sense for you.

## Other Social Media Platforms

Want to provide a way for committed constituents to connect? Social networking websites are free online communities where supporters can easily keep up to date on your organization and, in most cases, "talk" with you and other supporters. The most popular are **Facebook** and **Twitter**, but there are many others that focus on a particular niche of users or platform features. This is how they compete on the market—and exactly why they may be useful to your organization.

There are hundreds of more specifically targeted niche social networks to choose from. Here are a few of the most popular:

- Instagram, which is owned by Facebook, is a photo and video sharing app that has become one of the most popular and engaging social media sites—especially for organizations that create a lot of visual content.
- Snapchat and WhatsApp are two messaging apps for mobile that are gaining in popularity and traffic, especially among teens and 20-somethings.
- Care2 and Brigade are platforms where activists who care about specific issues can connect and work together.
- Nextdoor, Yik Yak, and Blasterous are for networking or finding information within a specific neighborhood or local community.
- Tumblr is photo-heavy and streamlined blogging platform that allows you to create a continuous stream of images and gifs.
- Reddit users submit links of the most interesting content on the web and others can then choose to "upvote" or "downvote" the content to determine how much priority the link is given.

# Websites

**Your website is likely the first place someone goes when they want to find out more about your organization. But it takes a bit of work to make a good impression.**

An intuitive web *Content Management System* can make it easy for staff members to post important information or calendar updates. It can also help you create interesting and *SEO*-friendly *Website Content* that includes useful tips, in-depth information, images, videos, or links.

Your website can also be an important part of your fundraising campaign. A good *Landing Page* or *Microsite* can stand apart from your regular content and focus a web visitor's attention on a specific action.

## Content Management Systems

Web Content Management Systems (CMSs) let you create and maintain customized websites, update their graphic design and navigation over time, take advantage of contributed modules, and automate routine updates—for instance, removing events from your homepage after they've come and gone. They allow staff members to update site content and navigation without technical know-how or web design experience.

Most CMSs won't let you update existing sites that were built in other systems, so if you want to implement a CMS for an existing site, you will have to rebuild, and sometimes redesign, the site. Many organizations hire consultants to build the initial site in a CMS, and then use the system to maintain it.

Widely used open source options include WordPress, Joomla, Drupal, and Plone. These systems are free to download, but you'll need someone with technical skills to set them up. Other systems, such as CommonSpot by PaperThin, Sitecore, and CrownPeak, provide sophisticated CMS functionality that you might be able to implement for $10,000 for the first year. Some systems, such as eZ Platform, Evoq from DotNetNuke, and ExpressionEngine, offer the flexibility of an open source system at a low licensing fee, or offer both a paid enterprise-level and free community versions.

For more straightforward sites, consider simple tools such as Wix, Weebly, or Squarespace that let you define navigation, pick a graphic design template, and enter text and images on simple web-based forms.

Larger nonprofits looking for both enterprise-level functionality and strong integration with constituent management may want to look into Blackbaud's NetCommunity or Luminate CMS. Some integrated online systems may also provide CMS functionality in addition to such features as constituent management and **Broadcast Email** functionality. In fact, many of the traditional CMS platforms mentioned above offer these services at additional cost.

## What About Updating Existing Sites?

Other desktop tools, like Adobe Contribute or Adobe Dreamweaver, also let less-technical people update websites, but they work by directly changing the code for individual pages. These are the only tools that will help update existing sites—however, they limit your ability to make updates that are more substantial or affect multiple pages. CMS systems are the better approach if you're building a site from scratch.

## Landing Pages and Microsites

Whether you're reaching out to supporters via email, social media, or direct mail, you need a next step—a place where you can provide information, help them take action, or ask for donations. You need a landing page or microsite.

Landing pages and Microsites often work well as the hub of a campaign because they can be streamlined to focus attention on specific information or actions. However, they're not interchangeable.

Landing Pages are meant to be simple. They're just a page within your existing Website that you send people to using **Broadcast Emails** or **Social Media** posts—for example, as part of a fundraising campaign. Usually there's an image and short text and a big emphasis on the call to action. For example, if your Landing Page is supposed to bring in donations, you don't want to distract supporters with a lot of information. You want it to be easy to access the donation form and a big "donate" button so that it's clear what they need to do.

Microsites are like miniature websites. They usually have multiple pages that follow a particular theme or topic. One advantage of a microsite is that it doesn't have to look like your organization's website. It can use different colors or a different layout, although it would probably be a good idea to include your organization's logo somewhere on the site. By using a separate design, you can make the site standout—for example, to distinguish a program or service or campaign from everything else your organization offers. Microsites are best for providing a lot of information, but they're also flexible enough that you can include secondary actions such as a donation or advocacy appeal.

## Online Advertising

Online Advertising lets you advertise your organization or mission through websites other than your own. Some sites limit you to text ads, while others let you run display ads or banner ads— usually an image limited to a particular size, though animation and interactive banner ads are sometimes possible. You typically pay for ads by "impression"—the number of times the ad shows on a site—or by "click"—the number of times any user clicks on the ad.

Google AdWords is a common, cost-effective method. You create a short text ad, choose the keywords and geographic area you'd like to reach, and Google posts your ad next to searches for those keywords. Google provides easy-to-use tools to track your results and further optimize campaigns, making it straightforward to manage. Cost depends on the popularity of the keywords you choose, but often starts at just a few cents for each user who clicks through to your site—and you can cap the amount you spend per day. Even better, qualifying nonprofits can get a huge number of free ads per month through the Google Ad Grants program.

*Social Media* sites, including *Facebook*, *LinkedIn*, and *Twitter*, also support online ads, usually for a similar price. Advertisers have the option to create an ad or "boost" a post and pay a set amount per impression or a variable one. They also let you target a number of different demographics. Facebook charges a $1 per day minimum in the U.S.

Many blogs and *Websites* also accept ads. These are a good way to target a particular niche audience. Companies such as Blogads and BuySellAds facilitate advertising across a number of different sites. These networks typically let you search for blogs and websites by demographic, audience size, and prices. Prices vary greatly depending on the blog, placement, and duration, but start as low as $1 per site per 1,000 impressions for less-known sites.

## Search Engine Optimization

Search Engine Optimization (SEO) isn't a type of software but a set of techniques to help search engines such as Google or Bing find your website and show it high up on the list of results people see when they search for terms you specify.

Many factors contribute to how high your website appears in search rankings and some of them are not clearly defined so that malicious sites (or eager startups) can't easily figure out the rules and begin appearing at the top of the search results.

First, encourage as many sites as possible to link to you. These links help search engines find your site, and the more incoming links from credible organizations, the higher you're listed in search results.

Second, identify keywords for which you'd like to be found, and use them prominently—for instance, to be found by those searching for "food pantries in Cincinnati," use the words "food," "pantry," and "Cincinnati" often in page text, prominent headers, titles, and even page file names.

It's more important than ever to have a **Mobile-Friendly Website**. Google now factors mobile friendliness in its search rankings. Other important factors include making sure you have title tags, a meta description, images, content that's at least 500 words long, and useful links.

A good web **Content Management System** can help place your keywords effectively, as well as help with the more technical aspects of SEO. **Website Analytics** tools can then help you track the keywords used to find your site.

It's often useful to go beyond SEO and consider paid placement. For example, some search engines let you buy ads for particular search terms. In particular, Google offers a Google Grants program that provides free Google Ads to qualifying nonprofits.

## Website Content

The blog was created as a way for individuals to have a website where they could post commentary, stories, links, or even photos, videos, audio files, or maps. Posts are shown in order by date, starting with the most recent. Subscribers can comment on what you've written, helping you to interact with your constituents and hear what's important to them. Organization websites quickly adopted the blogging style and ethos and now nearly every nonprofit offers a blog section.

However, you don't have to call your website content a "blog" or restrict your commentary or multimedia content to a particular section. A useful and interesting website will offer a variety of content, update the content multiple times per week, and prioritize the needs of your website's visitors.

A number of software packages make creating and updating website content quick and easy, even for non-technical users. Many web **Content Management Systems** offer blogging functionality—if you're using one to update your website, it almost certainly makes sense to start there. If you need to look further, Blogger, WordPress.com, and Typepad are all commonly used and very affordable. Tumblr is a free and customizable option, and is especially useful if you are posting a lot of visual content, but you cannot host it on your organization's website. Medium is great for magazine-style blogging. WordPress.org and ExpressionEngine offer more advanced functionality for those with some technical expertise.

All of these tools let you set up a blog, customize its appearance, add links to important actions such as donating or signing up for a newsletter, and then post text, photos, or videos. All support reader comments and let you moderate comments for inappropriate content or spam. These tools also let readers subscribe to your blog via RSS tools—an important functionality, since some people read blogs solely via RSS.

Most tools also support multiple bloggers, which can help share the expertise of your staff or volunteers and help keep the blog fresh and interesting. More sophisticated blog tools such as WordPress.org and Movable Type, and web **Content Management Systems** such as Drupal let you manage complex blogging workflows—for instance, setting a central administrator who approves posts written by multiple bloggers—or integrate blog posts into larger websites in sophisticated ways.

# Video Sharing and Streaming

Videos can provide a compelling way to tell your story online. While professional looking video can be expensive to produce, both in staff time and actual money, Video-Sharing websites let you upload videos to the web for free. Once they're online, viewers can comment and share them with friends. In general, you maintain ownership of the videos you post, but you grant the site certain rights. Before you post a video, read the site's policies and terms carefully.

There are many free Video-Sharing options, including YouTube, Dailymotion, and Vimeo. Brightcove lets you show videos and video pages without any logo or branding for Brightcove itself, but starts at $99 per month. YouTube offers nonprofit-specific functionality, such as the ability to create a branded YouTube channel, link calls-to-action directly in videos, and accept donations directly through the video page. It also provides video production space for nonprofits in a few cities around the world.

Live streaming video is becoming increasingly popular. *Facebook* Live lets you capture and stream video as it's happening. Periscope is a favorite option for *Twitter* users and YouTube offers its own live streaming capability. Wowza, UStream, and Livestream are three paid options that give you more features and the ability to save and edit the videos you created.

Many Video Sharing websites also allow you to post the videos on your web page or blog. They provide HTML (the coding language of websites) for you to copy and paste to embed the video. In most cases, the sites' logos are displayed on the embedded video players.

There are also mobile apps that allow you to create shareable videos right from your smartphone. Instagram, Cinemagram, and GifBoom offer tools to create, edit, and share short, simple video clips. While these applications have their limitations, they can be powerful tools for expressing a brief message that couldn't be easily communicated with a photograph.

# Webinars

Many people learn best in person, but when you can't get everyone together in the same room, Webinars are a great, cost-effective way for communicating a lot of information. Several useful Webinar platforms are mentioned in the *Online Conferencing* section of this guide. This section looks at what it takes to manage a Webinar successfully.

Your attendees need a **Landing Page or Microsite** where they can register and get additional information. When you set up your registration page, make sure to include a description of the session and the date and time when the event will take place. Many Webinar platforms will create these pages for you automatically. Before your registration page goes live you'll also have to decide whether you want to charge attendees a fee and what information about attendees you want to collect.

Next comes promotion. Think through how you'll get the word out about your webinar. A **Broadcast Email** can get the attention you need. **Social Media** is another good promotion channel, especially if your event is free, because it's likely your followers will share the link. You might also find that listservs or other communities are full of the kind of people you're hoping will attend your webinar.

It's common for people to register for your Webinar and immediately forget about it and lose track of all the details. Make sure to send an auto-response email with access information and a receipt (for paid Webinars. Then send at least one reminder email in the week before the Webinar. Some organizations send an additional reminder an hour or two before the Webinar begins. Again, some platforms, such as GotoWebinar, will automatically do this for you.

Interactivity is an important element of any webinar. At the very least, you should make it easy for people to ask questions—but make sure that you're good at muting and controlling background noise if people ask questions out loud. Another option is to have them ask questions using the conferencing tool's chat feature, if available. Many *Online Conferencing* platforms also allow you to create polls that show you and your participants real-time results.

Once your Webinar is finished, there are still ways you can enhance the experience. Before you log off, make sure you capture chat or whiteboard activity. Also, it's a good idea to get feedback via a post-Webinar survey. Later in the day, send attendees useful links or resources. These can include a recording of the Webinar, articles or links discussed during the session, the survey link, and information about related Webinars or resources.

*To put technology to use for your organization, you'll need to choose the right tool and implement it—often a daunting process. Complex software can be expensive, and implementation can be difficult. You also need to plan for training and supporting staff and moving your data to the new system.*

*By following the steps outlined in this section, you can navigate this process successfully, remove the frustration, and minimize the risk. You can also learn to use cost benefit analysis to help you prioritize which technology projects to tackle on a limited budget.*

CHOOSING AND IMPLEMENTING SOFTWARE

# Choosing and Implementing

**Before you set out to select and implement a new piece of software, you need to determine which projects are worth pursuing—and how to prioritize when you have more needs than budget to fund them. A cost benefit analysis can be an effective means of Prioritizing Technology Projects.**

Once you've got a list of projects to pursue, it's not enough to know what kinds of software might work for you. You also need to carefully think through how to find the right software for your needs, and how to get it up and running smoothly.

Regardless of the type of software you're considering, the first critical step is *Defining Your Software Needs*. For a minor purchase, this step might involve a quick conversation with other staff, but for a large, mission-critical piece of software, it might take months of work. Before you go any further, make sure you don't already have a system that will do what you need. Can you expand an existing system to take on more functionality? Making use of existing systems means saving the time, money, and frustration of choosing, installing, and learning a whole new application.

If you do need new software, the next step is *Creating a Software Shortlist* to winnow down all the possible options to a manageable list. For a minor purchase, this might mean simply talking to a few people and choosing a single package to explore further, but if you're making more of an investment, you'll want to investigate more in-depth and identify a list of three-to-five software options.

The next step is *Evaluating Software*—you wouldn't buy a car without test-driving it, would you? Try out each system on your own, or ask the vendors to demo them for you. Make sure to evaluate them for the real world scenarios for which you'll use them and not just a hypothetical list of features.

As you're considering different options, you may find yourself **Comparing Open Source and Proprietary Software**, or **Comparing Installed and Cloud-Based Software**. There are pros and cons to each of these different types, but what matters most at the end of the day is the ability of the tool you choose to meet your needs and the cost of the system both upfront and over time. **Evaluating the Software Contract** is critical to making sure you know what you would be purchasing—especially for more complicated software.

If you're choosing a complex, mission-critical component, consider hiring a consultant who knows the market and can help define your needs. **Choosing a Consultant** doesn't have to be scary—think of it as a way to add valuable outside experience to your team.

Once you've chosen your software, you're only about half-done— you still need to implement it. Depending on the type of system you've chosen, you may need to think about **Migrating Data**, or moving it from your old systems into your new one. This is rarely an easy step, and it requires careful consideration and planning. In addition, no matter how amazing your new system is, it's useless to you if no one knows how to use it—however big or small your new system is, make sure you plan for **Training and Supporting Staff**. Who should they turn to with questions? What should they do, or not do, with the system?

No system will maintain itself—particularly one that includes data. **Caring for Your Data** means establishing policies to ensure your data stays clean and actionable, and that it's easy to access the information you need from the system. The best way to keep data useful is to do so right from the start: What should staff think about when entering records? Who will monitor data quality? Help your staff know what they should enter and when, and define the steps that will ensure your data is clean and usable when someone tries to find something.

# Case Study: Implementing Software

How can a nonprofit most effectively implement software? We've provided an example of how a fictional-but-realistic organization used best practices to introduce new technologies. All of the processes highlighted within the text are covered in more detail in the next section.

## Hacer

*Putting Technology Into Place*
*$250,000 Budget*

Hacer is a small, expanding organization with a mission to advocate for the rights of immigrants and farmworkers in Oregon. Until now, the staff has used Excel for everything from donor-tracking and budgeting to human resources. Gabe, the executive director, decided the organization has outgrown the tool, and applied for and received a grant to cover the cost of a database—the largest single purchase in the organization's history.

To begin with, Gabe sat down with the staff for a conversation about **Defining Software Needs** and identifying the many goals they wanted the new software to accomplish: tracking donors and the army of volunteers that carry out the nonprofit's organizing work, tracking interactions with state and national legislators, coordinating broadcast emails and the quarterly newsletter to keep constituents up-to-speed on immigrant harassment issues and relevant legislation, and managing the fundraising events that support both the staff and the mission. While this process immediately yielded a long list of "must-have" features, subsequent meetings culled the list to those features most necessary to the organization's day-to-day functions and an additional list of desirable "nice-to-haves" that were not deal-breakers.

Since Hacer would be working with a system for the first time, Gabe asked around at other nonprofits about **Choosing a Consultant** to help the staff understand and streamline their processes, and hired someone to make sure they were getting off on the right foot. The consultant talked Gabe through the various options, including *Comparing Open Source and Proprietary Software* and *Comparing Installed and Cloud-Based Software*, and helped *Create a Software Shortlist* by identifying the most-common software tools used by other advocacy nonprofits with similar needs. Based on her own experience, she made recommendations about which might be a good fit for Hacer. Of her list of five, two were out of reach financially, leaving a final list of three.

Gabe scheduled vendor demos to facilitate *Evaluating the Software* on the list, and sat down with the consultant and the key staff members who would be using the system. Most of the systems more than met its needs—both current and anticipated as it continues to grow—but the demos made it very clear that one system was substantially more user-friendly for what they needed to do. Nick worked with the vendor to negotiate terms, asked the organization's attorney to *Evaluate the Software Contract*, and made the purchase.

The next two months passed quickly. Because all Hacer's data was spread out across a number of Excel workbooks, staff worked with the vendor and consultant on *Migrating Data* into the new database to make sure it was consolidated and secure. The vendor also worked toward *Training and Supporting Staff* on using the system and *Caring for Data*, including teaching them the best data-entry and recordkeeping practices to ensure that Hacer would get the full value of what the new software could bring to the organization.

# Prioritizing Technology Projects

Given unlimited budget, unlimited staff support, and unlimited time, you wouldn't have to prioritize your technology projects. You could do them all. Right now. But we don't live in that alternate universe.

Here, in the real world, we often have to make tough choices when it comes to technology—especially if the basics (reliable computers, email, calendaring, file sharing, and backup systems) are already in place. There may be four or five solutions that have the potential to leap your organization into the future. How do you go about choosing?

## Start with a Cost Benefit Analysis

It's not a new method, but it's a tried-and-true one. Weighing the benefits vs. the cost can save you a lot of time, money, and frustration. On the benefit side, look at how much your new technology will increase your ability to raise money, to reach constituents, or serve your mission better. Also, consider whether it will save time or money by enabling you to work more efficiently.

Of course, there are dozens of projects that might provide at least a little bit of benefit. You also have to consider the costs. How much money will you have to spend? How complex is the project? Will you need to bring in outside consultants or help?

Sorting out these and other questions can help you get a clearer picture of what project is going to provide the biggest payoff.

## Start Comparing

Using the Define Your Priority Projects worksheet, you'll answer five questions to come up with a score that will help you compare your projects and sort out what to get started on next. We'll walk through the five questions below.

# How Much Does it Increase Effectiveness?

Ask yourself, Will the project transform the way your organization is able to carry out its mission? Or is it simply an operations change that doesn't directly affect your mission?

For example, let's say you're considering combining two constituent databases—the database where you track events, and your main database. Let's say you're really struggling to make sure all the right people are invited to the events. You know you have a lot of clients who are interested in and would benefit from your events, but many of them fall through the cracks and don't get invites or follow-up emails. When you step back, the project is somewhat operational, but it could mean more people at events, which might make a difference for those new attendees and your mission as a whole. If we were to score this project on our worksheet, we might give it a six or seven out of 10. (We've filled in this example on the worksheet.)

# How Much Does it Reduce Time or Costs?

Time and money are always major factors in a technology decision. Ideally, the new technology will reduce both, thereby freeing up more resources to put toward your mission.

Returning to our database example, let's say two different teams are managing those two different databases, and they're both pretty happy with the current system. You will probably save time managing the data and in pulling lists, but it's not an enormous amount. Score it a four out of 10.

# Is it Straightforward to Implement?

Technology projects can be deceptively complex. Can you map out in your mind how you'd go about this? Do you know the steps you'd take? Does it seem as though it can be completed quickly? If you answered "yes" to each of these, then you'd give your project a high rating. (Note that we've changed the rating scale here. You're rating from zero to five.)

Let's look at this category in the context of our database consolidation example. Is this project straightforward? Well, you're certainly not going to be able to make it happen overnight. You'll need to think through how the event data maps into the main database and how people will use this new combined database. Even with a carefully detailed map, moving data can be complex. Most experts recommend a couple of test runs first to make sure all the bugs are worked out. Let's give this example a two.

## Is it Inexpensive?

Then there's the cost. Putting value aside and just thinking about your budget, is the project expensive? If so, it gets a zero as an "inexpensiveness" rating. However, if it's free, then it gets a five. If the software and equipment are free or already installed, as is the case with our database example, but you need to bring in a consultant because you don't have the necessary expertise on staff, then the total cost is definitely not free. We've scored that scenario a one for "inexpensiveness" on our worksheet.

## Can You Do it In-House?

Lastly, think about whether you'll need to bring someone in from outside the organization. For our database example, we already decided you'd probably have to bring in a consultant. Although you could probably struggle through the data conversion alone, that can be risky because getting it wrong could mean working with flawed data for months. However, you're more likely to be able to pull off data conversion than build a mobile responsive website or create your own app, so let's give it a three.

## Add it Up

Once you're done, just add all the scores together. In the example, our database project gets a 16 out of a possible 35.

What's important about this number is not the number itself, but how it allows you to compare other projects, so let's quickly run through another example. (We've marked this one on the worksheet, as well.)

Let's say that your printer is pretty unreliable at the moment and you're thinking about getting a new one. It goes down unexpectedly sometimes and some print jobs come out illegible. Will a new printer increase your effectiveness? A little, maybe. Staff members sometimes don't have documents when they need them because they can't print. Give it a two. Does it reduce time or costs? Not costs, but time. It's not a huge amount of time, but there's some savings there. Give this category a four. Is it straightforward to buy a new printer? You'll need to do some research, but buying a printer is not very complicated. Give it a four. Is it inexpensive? A new printer will cost between $200 and $400, which is not a lot of money when you consider its percentage of your annual budget. Give cost a five. Lastly, can you do it in-house? Installing and networking a printer requires very little technical skill or knowledge. Give this a five.

The printer scores a 20, which is higher than the database project, even thought it doesn't provide as much overall benefit. Why? Because its costs are so low. You get more bang for your buck.

This process is rough and inexact, so don't agonize over the precise scores for each column. The end goal is to get you to a place where you can prioritize one or two projects and see them through to completion, so if, after seeing your scores, you also want to factor in a "gut check," go for it. The worksheet isn't the boss of you. Feel free to disagree with it, as long as you know why you're doing it.

# Defining Your Software Needs

Before you invest in new software, first determine exactly what you need. If you're buying just an inexpensive piece of software for one person, defining your needs can be a quick process, but if you're buying a major piece of software, itemizing your requirements will require a serious commitment—and perhaps take a month or more.

It's critical to consult with your staff and representatives of anyone who will use the system about their desires, critical needs, and frustrations. Make a list of what will be useful to your organization. If you define more than a dozen or so needs, prioritize them—what's critical, and what's just nice to have? If a requirement is critical, that means you would discard a system from your list simply for not meeting it. When you don't prioritize your requirements, it's all too easy to hold out for a system that's ideal in every way—which isn't something you're likely to actually find, or afford on your budget.

Once you have a list of requirements, evaluate your current system against that list to make sure it truly won't work to meet your critical needs. It's worth calling the vendor to ask—you might not be aware of all the processes your current system is capable of. Switching systems often requires substantial time and cost in selecting, moving data, and training, so make sure you aren't wasting resources replacing software that would actually work fine for your needs. Investing in better training is often more cost-effective than switching entirely.

If you are having trouble understanding how a system can best help your organization, consider hiring a consultant who can help you see how your organization handles data, and offer solutions that have worked for other clients in similar situations. When the time comes to work with vendors and find the right software, you will be glad you can understand and explain your organization's needs with confidence.

# Creating a Software Shortlist

There sheer number of software companies out there will make your head spin. Some initial research can narrow down your list of choices to a more manageable size and allow you to conduct a reasonable comparison of vendors and systems.

Start by seeing if credible sources have already provided a vendor roundup as a starting point. Idealware (www.idealware.org), NTEN (www.nten.org), and TechSoup (www.techsoup.org) are all good places to get started, and organizations that support your particular sector—for instance, an association of nonprofits with a focus in your area—might also have useful resources.

Speak with other organizations about what software they're using and the overall effectiveness of their systems. This will give you a general overview of some of the big-name software you should be considering as well as perspective on how software can help your organization. Keep your needs in mind, and you should be well on your way to making an informed decision.

If you're only making a minor software purchase, it may be sufficient to identify one or two tools that seem like they will work. For a larger investment, create a list of three to six tools for deeper investigation.

In the past, sending out Requests for Proposals (RFPs) to software vendors was more commonplace, but in the current market, strong vendors—including those likely to be best for your needs—may choose not to take the substantial time required to answer an RFP when they don't know the likelihood that they'll win your business. If you have a very large project that will look particularly good in a vendor portfolio, or if your organization requires you to work with a conservative bidding process, sending out a shorter, more general Request for Information (RFI) might get the information you need without wasting your time and the vendors'.

For a costly purchase, consider seeking out a consultant to look at your organization's needs and offer software suggestions. Consultants often have a solid understanding of the current marketplace and can recommend the software packages most likely to fit your needs. An extra set of eyes can also be useful in determining software features your organization could use but may have overlooked. The added cost of hiring a consultant might be offset by purchasing a software package that will be right for your organization not just now, but as your needs evolve in the future as well.

# Evaluating Software

Once you have whittled down your list to the major contenders, it's time to begin working with vendors and demoing software. Vendors can help your mission reach new heights, or they can hold you back—it's important to take the time to learn everything you can about them, and about their software, before making a final decision.

Most vendors will be happy to provide a trial version or demo their product over the web, which makes it easy to see their systems in action. If a vendor is going to give you a guided tour, take some time beforehand to define the specific features and functions you want to see and send them to the vendor in advance. If you leave them to define their own tour of the product, you'll likely get a great view of the strengths that glosses over the weaknesses. They should be able to give you an overview of how the software meets the needs you care about—be cautious if a vendor seems to repeatedly misunderstand or talk around a question you've asked several times, and be sure to ask the vendor to clarify if you don't understand how something they've demoed meets the needs you asked about.

Don't get wrapped up in exploring features you may not need. Complex features and fancy graphics can seem engaging, especially when they're highlighted by the vendor, but you will learn more useful information about the software by running through the processes you most need. In the end, it doesn't matter what a system does if it doesn't do what you need it to. To make the most out of your software purchase, you should also research vendors for specific information regarding their longevity in the marketplace. For example, how long have they been in business? How many of their clients are organizations similar to yours? An online search can reveal a lot about your vendor's history, including major changes to the company, litigations, and profitability.

Your relationship with a software vendor is one your organization will maintain for several years, so take the time to find not only software that works for your organization, but also a vendor that clicks with you.

# Comparing Open Source and Proprietary Software

Many organizations have strong feelings when it comes to comparing open source software to proprietary, vendor-supported options. Some advocate open source products (those built and supported by a community and then given away for free) as totally customizable, feature packed, and completely free software. Others believe that having a company behind a product, including the support and implementation help it can provide, makes proprietary, vendor-supported software the better bet. In practice, there's a lot of blurring between open source and proprietary software in today's market, and it probably makes sense to compare the options based on your own needs rather than assuming one model or the other will work better for you.

There are specific differences between the models, however. When purchasing proprietary software, a vendor will typically assist you in setting up and tweaking it to your needs, but you'll need to pay a sometimes-substantial fee and sign a contract. With an open source package, you simply download the software for free—but any installation, training, or customization is left up to you or to a hired consultant. There's no contract with an open source tool; you can start or stop using it at any time, but no contract means no warranties. It can be more challenging for organizations to troubleshoot bugs and hold developers accountable for problems with open source software.

If your organization has never used open source software before, it may be difficult to get it working perfectly right away, especially without hiring a consultant. Historically, many open source tools have been designed around what makes sense to developers rather than users, making them sometimes more difficult to use. Vendors for large proprietary software packages often have staff members trained to work with organizations from the ground up, which can be helpful—but which may make you reliant upon the vendor rather than your pick of qualified consultants for help.

This applies to support as well. Vendors often provide straightforward support packages, but that's often your only option. You may need to devote more time to finding a consultant to support your open source implementation, but if you're not happy with one consultant, you can find another that better meets your needs. These days, in fact, some open source developers offer the same type of support, services, and even installation as a vendor, for a fee.

An open source tool by definition is open to modification by programmers, so you can make it do just about anything given sufficient time and money. Unless you're devoting a huge budget to your project, however, this probably isn't a practical path. Instead, look at what configuration options your software has, and what community-built add-ons are available. Open source tools typically are strong in both elements, but some specific proprietary tools are equally strong.

In general, your final choice shouldn't be based on whether a system is open source or proprietary, but on how well the features fit your needs and the total cost of the product. Many proprietary software vendors will provide free trials for their software so you can try them out alongside their open source competitors. Consider the cost of the time involved in setting up and learning any new software as well as the technical knowledge needed to maintain it.

# Comparing Installed and Cloud-Based Software

Cloud-based software goes by many names: Hosted software, Software as a Service (SaaS), ASP Software, and On-Demand Software. Such systems can offer a variety of benefits, including easy maintenance and remote access, but give users less control over updates and require a reliable internet connection.

Typically, software used by an entire organization is purchased and installed on a server or on many computers at once. With Cloud-based software, you pay a monthly fee, and any licensed member of your organization can access the software from anywhere with a speedy internet connection—in most cases, even on smartphones and tablets. Many respected software companies are making the switch to Cloud-based options, and everything from online payment processing tools to full-blown donor management systems can be accessed this way.

One of the largest misconceptions surrounding Cloud-based software is the lack of security. Since your data isn't stored on your computer, it's more liable to threats, right? Not at all—big software companies have the resources to back up your data, keep your software up to date, and protect you against threats often more effectively than a small organization could. Still, the company does have access to your data, and if you work with data that you might need to subpoena for—like case records for your work with Iraqi immigrants—an installed system might be the safest bet.

Most Cloud-based software is paid for with a monthly or yearly subscription. It's similar to a "rent" vs. "own" model—it's likely to cost less upfront to rent Cloud-based software, but the monthly subscription to use the software can add up over time. If you find you need to store a lot of data and allow many users to access it, you may be grateful you paid the one-time fee for installed software a few years down the line. Keep in mind, however, that you will also need to pay to buy and maintain the servers used to run an installed software.

Being online, Cloud-based software can frequently integrate seamlessly with Broadcast Email, online donation tools, and websites. Additionally, most Cloud-based software can work from anywhere—including from home or on the road—and can be accessed equally well with Windows, OSX, and Linux, provided you are using a compatible web browser.

Much of the work of a system administrator is left to the vendor with a Cloud-based system. The vendor installs updates, maintains the servers, and monitors the system to keep it up and running so you don't have to. If you don't have any IT staff, this can be a huge benefit. However, this reduces the amount of control you have over the software. For example, the vendor may automatically roll out new features that might confuse your users.

In general, choose your software based not on whether it's Cloud-based or installed, but on features and the total cost of the product. Weigh the desire to save money in the short term, and the need to pay someone to maintain an installed system, with the possible long-term cost savings of an installed system in the long run. And make sure you compare the software package carefully to your list of requirements—it doesn't matter if the software is installed or in the Cloud if it doesn't do what you need it to do.

# IT Security Policies

Most staff members want to keep your organization safe. But they often simply don't know what is acceptable and what isn't, and that can easily create security problems. Clear policies and procedures are an important step toward building awareness and strengthening habits.

What should a policy include? It should outline dos and don'ts in as much detail as possible while maintaining some flexibility to address new threats as they emerge. A typical policy might include an explanation of the policy, an outline of who is subject to it, and a clear description of who is responsible for carrying it out. You might also need to define particular terms or commonly misunderstood concepts, and you'll definitely need to state the consequences for not following the policy.

Different kinds of written policies to consider include:

**Acceptable Use Policy:** This stipulates how staffers can use information, electronic devices and computers, and network resources. It's typically a broad policy that tries to account for bad behavior on your devices or network.

**Bring Your Own Device/Mobile Device Policy:** This outlines mobile device procurement and/or reimbursement. It also can outline requirements for how staffers can or cannot use personal devices.

**Employee Internet Usage Policy:** Specific to the internet, this includes web browsing, instant messaging, file transfer, file sharing, and other standard and proprietary protocols.

**Email Usage Policy:** This covers appropriate use of any email sent from an organization email address and applies to all employees, vendors, and agents operating on behalf of your organization.

**Email Retention Policy:** This is intended to help employees determine what information sent or received by email should be saved and for how long.

**Password Construction Guidelines:** This provides best practices or guidance that will help staff members create strong passwords.

**Password Protection Policy:** Rather than a guideline, this policy establishes a minimum standard for the creation of strong passwords, the protection of those passwords, and the frequency of change.

**Mobile Employee Responsibility Policy:** This policy describes the requirements for staffers who work outside of an office setting.

**Social Engineering Awareness Policy:** This is intended to make employees aware that social engineering attacks occur, and gives employees specific procedures for dealing with social engineering attempts.

**Software Installation Policy:** This outlines what kinds of software can be downloaded and what cannot.

**Server Monitoring Policy:** This is designed to protect the organization against loss of service by providing minimum requirements for monitoring servers.

**Server Audit Policy:** This is meant to ensure all servers are configured according to the organization's security policies and to set a minimum requirement for how frequently servers are audited.

**Security Response Plan Policy:** This establishes the requirement that all business units supported by the IT Department develop and maintain a security response plan. Many also outline the requirements for such a plan.

**Social Media Usage Policy:** This establishes guidelines for proper usage of social media outlets.

# Choosing a Consultant

Consultants can offer your organization a wide range of benefits. Perhaps you need a fresh perspective, some good advice, or just a little extra help during a busy time of year. If you understand your goals and determine what exactly is beyond your capabilities, an outside voice can help your organization surpass what it could do alone.

Your organization can gain a lot of valuable knowledge and first-hand experience from working with a consultant. If you are looking at improving your donor management system, for example, a consultant can impartially identify the weak points of your current setup and give you examples of how similar organizations have dealt with those problems. If your staff knows what they need but can't decide on a solution, a consultant can help define a process to weigh pros and cons and cut through internal politics. If you're heavily customizing a system or building one from scratch, bringing in a consultant will almost be a certainty.

To find a good consultant, start by asking other organizations for recommendations. State nonprofit associations, relevant email discussion lists, or groups that are focused on your cause can be other good sources for potential consultants.

Schedule a phone call with a number of consultants. Talk to them about the goals of your project, and what you envision as an end result, and ask them how they would achieve those objectives. They should be able to talk about one or more ways they could get to that end result and not just sell you on the big picture. Make sure you can understand what they're telling you—a consultant who can't speak language you understand is a red flag. Since your consultant will be a valuable member of your team, treat the interview and screening process seriously, and check references to hear about other people's experiences.

# Evaluating Software Contracts

A contract with a software vendor can be a complicated document with terms and conditions you might not immediately understand. If you're purchasing a small piece of software from a vendor who deals with tens of thousands of clients, you may not be able to negotiate your contract, but you can always choose to walk away. There are some basic concepts you should be familiar with to ensure that your contract meets your needs, and not just the vendor's.

You should make sure that your vendor has included what course of action should to be taken if you have a problem. If the vendor has promised anything to you in the past, it must be included in the contract. For example, if the vendor promised that your software can be fully integrated with your Broadcast Email tool, and you find it doesn't live up to your expectations, the vendor should be legally obligated to work with you to find a solution. Additionally, you should make certain there are fair penalties for both you and your vendor in the event of small or major issues. If you cancel a consultation, for example, make sure that the fees for doing so are within reason, and that there is some recourse should they put you in the same position.

The software license is a major component of the contract. There should be no language implying that your access to data is restricted, or that your data can be used by the vendor in any way—outside of showing you your data inside the software. On the other hand, you should be able to do what you need with your data. If you are concerned that your use of the software may void the warranty, ask to have it changed. For a larger software purchase, contract negotiation is common and sometimes even expected. If the vendor is performing consulting work with your organization, the Scope of Work (SOW) is often an attachment to the contract that outlines what is being purchased, by whom, and for how much. It can also include a timeline for implementation of the software, or any other tasks to be completed by the vendor. In general, the more specific these outlines are the better. Especially in the case of a large software purchase, you may want to consult with a contract attorney.

# Migrating Data

Moving data from one system to another, typically called "data migration," can be a long process. If you need to move complex data like accounting records, constituent data, or web pages, you'll need someone with significant experience working with databases to ensure a smooth transition. It's very rare for the data in one system to easily map into a new one; most of the time, there are complex considerations of where, precisely, to put all the information in your new system, and how to get it there. Most organizations will benefit from hiring a consultant for this process.

The length of time it will take to transfer your data is highly dependent on its current condition. If your organization has been keeping records for years without carefully managing them, your data may be out of date and full of erroneous and duplicated information, and pulling together and cleaning the data could take much longer than the process of actually moving the data. It's critical to clean the data before starting to use your new system. If this step is skipped, problems caused by bad data can heavily affect your staff's comfort with a new system. Cleaning the data in your old system is likely to be a much better plan than trying to fix it in your new one.

If your data is particularly messy, or the process of migration particularly complicated, consider carrying over only the most important data. While this clearly isn't ideal, it might boost confidence in your new database and save time in the long run to move only the data that you're sure is accurate and fill in the blanks over time when new information needs to be added. Some organizations, in fact, choose to keep both their old and their new systems running in parallel for a limited time, ensuring the old data will still be accessible until you no longer need it.

# Seven Data Migration Milestones

From start to finish, these are the basic steps to a smooth and efficient conversion process.

1. **Talk to a consultant.** Even a short consultation can help you understand your organization's unique needs and put you on the right track.

2. **Think through the logistics.** Assigning an experienced staffer to own the project and the data is essential for keeping the project on time and on budget.

3. **Clean your data.** You don't have to do it all up front (there are many post-conversion tools to help with cleaning), but the more you do before the conversion the more useful your new system will be after.

4. **Create policies and procedures.** Unless you document rules for what goes into the system and how it's entered, you're in danger of creating a system full of cumbersome and confusing data.

5. **Develop a map that shows how data moves from your old system to your new one.** This is the surest way to make sure your data is usable in your new system.

6. **Test your conversion.** A conversion is rarely perfect on the first try. Or the second. Leaving plenty of time to test and adjust will mean less pressure on your conversion specialists and staff members to get the job done and more chances to make sure it's done right.

7. **Convert your data.** If planned properly, this step is a matter of handing off your data and going live with your new system.

If you're planning to hire a consultant to transfer your data—which is often desirable—it's generally more valuable for them to thoroughly understand the ins and outs of your new system rather than your old one. Your new vendor can often suggest a good consultant for your needs. In working with them, try to be as specific as possible about your goals and expectations, and allow plenty of time for the switch to happen. If problems arise, new solutions and compromises may need to be agreed upon with the consultant.

Whatever method you choose, make sure that you do some trial runs of your data conversion before moving the data, and some serious spot checking of the data in its new location to ensure that data migration will be as painless as possible.

# Training and Supporting Staff

Training and support are one of the most important, but frequently overlooked, steps of implementing new software. It won't matter if you buy the most sophisticated software on the market if your staff doesn't know how to use it. Keeping your entire organization up to date will help to ensure that you are making the most out of your purchase.

To design your training, assess what you need from your system, including how it will affect each of your staff members and what processes will change. For example, is it worthwhile to train everyone in your office on how to enter event information into the system if you only have one event per year? Who is assigned to enter data, and how should it be entered? Who will need to know if something goes wrong?

When signing a contract for a new system, discuss how your nonprofit will be trained, and how much it will cost. Your software company will often have a training package to help you get to know the system inside-and-out. Alternatively, independent consultants can provide useful training that's often more tailored to your specific needs.

Don't plan to just train your staff once and assume they'll be experts forever. Regular refresher courses are important to maximizing productivity with your system. A lot of software designed for nonprofits can be complicated and packed with features not everyone will use daily. If you have many volunteers who work with your system, or a high rate of employee turnover, it's even more important to stay on top of training. Having a quick reference for the features your organization uses the most can be helpful for the entire staff.

In addition, it's important to know who your staff can turn to if they have a problem. Software vendors will usually offer support over the phone, by email, or over Skype—however, you shouldn't have to wait for customer service every time someone forgets how to print a report. If you don't have a dedicated technology guru, someone on your staff should be the one to go to with little problems, so things can keep moving smoothly. Consultants can also be helpful, especially if vendor-provided support is limited or non-existent.

# Caring for Your Data

To get the most out of your data, it's important to keep it organized. Many nonprofits suffer from unwieldy lists of donors, volunteers, and even accounting records. A little process refinement can make your data much more accessible and effective while making your life easier.

You should establish specific rules as to how to enter new data. If one volunteer enters "Mrs." into the Title field, and another enters "Marketing Manager," it can be difficult to find what you need. Identify what information your organization absolutely needs for each record—for instance, is a record that only has an email address and no other information valuable, or should that person not be entered until there's more information?

It's also important to define the timeframe for entry—should a check be entered into the database within an hour? A day? A month? These kinds of rules not only keep your data clean, but ensure that everyone is on the same page as to what to expect from the data. Once you define the rules, make sure you let everyone know what's expected. This will likely require not just a one-off training, but periodic refreshers.

Making sure you know who can access and enter data will help as well. It's critical to ensure that one person is accountable for the quality of your data, but you'll likely want multiple people to be able to access it. It's a balancing act to ensure that you allow access to the people who really need it, but not so many that there's a lot of possibility of error or duplication.

It's also critical to set up a periodic process to check in on and clean your data. In addition to checking for duplicates and errors, outdated information may need to be deleted. Define when a constituent record is no longer useful to your organization—for instance, if a constituent is deceased or hasn't been active in years, should you delete or archive them? An unwieldy system full of everyone your organization has ever interacted with over 50 years is going to slow down your staff's ability to quickly get what they need.

*Want to learn more? In this section you'll find a list of websites where you can get more help, including the websites of nonprofits who provide software discounts, free publications, and membership support to other organizations.*

FOR MORE INFORMATION

**idealware** Idealware www.idealware.org

Idealware.org offers hundreds of free articles and reports about all the nonprofit technology topics that can help organizations better meet their missions, as well as a detailed archive of training.

**techsoup**.org the place for nonprofits and libraries **TechSoup** www.techsoup.org

TechSoup.org offers nonprofits a one-stop resource for technology needs by providing free information, resources, and discounted software. They provide instructional articles and worksheets for nonprofit staff members who make use of information technologies, as well as technology planning information for executives and other decision makers. In addition, their TechSoup Stock program offers more than 600 donated and discounted products at very low administrative fees.

**NTEN** Nonprofit Technology Network **NTEN** www.nten.org

NTEN is the membership organization of nonprofit professionals who put technology to use for their causes. They bring together a community of peers who share technology solutions across the sector and support each other's work. They enable members to embrace advances in technology through knowledge-sharing, trainings, research, and industry analysis.

# About Idealware

Idealware, a 501(c)(3) nonprofit, provides thoroughly researched, impartial, and accessible resources about software to help nonprofits and the philanthropic sector make smart software decisions. By synthesizing vast amounts of original research into credible and approachable information, Idealware helps organizations make the most of their time and financial resources. Visit www.idealware.org to learn more or view our hundreds of free articles, resources, and reports.

# Authors and Reviewers

The Field Guide is the product of the collaboration of dozens of people in all areas of nonprofit technology. Each edition builds on the content of previous editions. This edition's authors and editors included the following people:

### Chris Bernard / Research and Editorial Director, Idealware

Chris is a career writer and journalist with two decades of experience in newspapers, magazines, advertising, corporate and nonprofit marketing and communications, and freelance writing. Prior to Idealware, he was managing editor of a newspaper and a senior copywriter at an ad agency. For the past seven years, he's overseen Idealware's editorial and communications efforts, driving the creation and publication of more than a hundred articles, reports, and other resources and managing the communications calendar. Outside of his work at Idealware, he's an award-winning author and a frequent speaker and lecturer at literary conferences and festivals around the country.

### Dan Rivas / Managing Writer, Idealware

Dan is a versatile writer and editor who specializes in translating complex information into compelling stories. Prior to Idealware, he was a copywriter and editor at a marketing agency that serves large technology and financial services companies. He also has experience as a freelance writer and journalist, a census enumerator, a bookseller, and a college instructor. He is a graduate of Willamette University and the University of Michigan, where he studied anthropology and creative writing.

## Matthew Burnett / Director, Immigration Advocates Network

Before joining IAN Matthew worked representing low-income immigrants at East Bay Sanctuary Covenant and Northwest Immigrant Rights Project, and served as a law clerk to Justice Z.M. Yacoob of the Constitutional Court of South Africa. He received his B.A. cum laude from the University of Washington, where he was a Mary Gates Scholar and elected to Phi Beta Kappa, and his J.D. cum laude from Seattle University School of Law, where he was a Public Interest Law Foundation grant recipient and on the editorial board of the Journal for Social Justice. In 2013 Matthew was named to the Fastcase 50, which honors the law's "most courageous innovators, techies, visionaries, & leaders."

## Gordon Mayer / Contract Researcher, Idealware

Gordon is a writer and storyteller who has been ensuring all stakeholders have voice in shaping effective and fair policy for more than 20 years. For the past decade, he's been leading advocacy, leadership development and capacity-building organizations as a development and communications director at Gamaliel Foundation operations director at People's Action, and communications coach and trainer at Public Narrative.

**Some existing content was also written for previous editions by the following authors:** Laura S. Quinn, Kyle Andrei, Andrea Berry, and Elizabeth Pope.

**This work would not be possible without the generosity of the following content reviewers:**

Kyle Crawford, *Fundraising Genius*

Linda Widdop, *Director of Technology Services, TechImpact*

Alice Aguilar, *Executive Director, Progressive Technology Project*

Steve Egyhazi, *Principal Multimedia Analyst, Indiana University*

Ian Pajer-Rogers, *Communications Director, Interfaith Worker Justice*

Alice Aguilar, *Executive Director, Progressive Technology Project (vendor)*

Jim Lynch, *TechSoup*

Nate Nasralla, *Director, Capacity Building Programs & Fundraising Solutions, Network for Good (vendor)*

Kathleen Pequeño, *Senior Account Manager, Fission Strategy*

Shana Glickfield, *Beekeeper Group*

Lisa Colton, *Darim Online*

Andrew Means, *Uptake*

Debra Askanase, *Community Organizer 2.0*

Peg Giffels, *Clarified Concepts*

Michelle Morel, *Morel Communications*

Daniel Chiat, *Measuring Success*

Glenn Rawdon, *Legal Services Corporation*

Maddie Grant, *SocialFish*

Andrew Cutting, *The Foraker Group*

Brian Yacker, *YH Advisors*

Rick Cohen, *National Council of Nonprofits*

Ed Mulherin, *eCratchit*

Roger Hagedorn, *City of Minneapolis*

Kelly Matthews, *New York Council of Nonprofits*

Larry Valez, *SINU*

Jennifer Chandler, *National Council of Nonprofits*

Brian Darby, *Consultant*

Lisa Rau, *Confluence Corporation*

Nomi Adler Dancis, *Independent Consultant*

Made in the USA
Columbia, SC
10 May 2017